A FRAMEWORK FOR THE FUTURE

GOVERNING ARTIFICIAL INTELLIGENCE TO SAFEGUARD SOCIETY AND NATIONAL SECURITY

First Edition

ISBN 978-1-949432-13-8

Any trade names, trademarks, service marks, etc., mentioned in this publication are for identification only. Therefore, any specific company or product mentioned is owned by their respective owner and not by Inner Alchemy's Publishing. Further, the company or product mentioned neither owns, endorses, nor has heard of Inner Alchemy's Publishing. By stating this, we can avoid printing the ®, TM, ©, etc. marks that we might otherwise have to place throughout the text.

The publisher does not participate in, endorse, or have any authority or responsibility concerning private business transactions between our authors and the public.

Published by:

Inner Alchemy's Publishing

332 S. Michigan Ave.
Ste 121-C141
Chicago, IL 60604-4434

Printed in the United States of America

TABLE OF CONTENTS

MESSAGE FROM THE AUTHOR

As artificial intelligence (A.I.) continues to reshape the world, it is crucial to approach its development and governance with both foresight and responsibility. We are witnessing unprecedented advancements in technology, with A.I. permeating every facet of our lives—from healthcare and finance to social media and education. These innovations bring tremendous opportunities, but they also present complex ethical, social, and economic challenges that must be addressed with care and intention.

The purpose of this framework is to guide policymakers, corporate leaders, and individuals as they navigate the rapidly evolving landscape of A.I. Our goal is to create a structured, ethical foundation for A.I. that ensures this technology serves humanity and respects the values we hold dear. This document advocates for a balanced approach, one that embraces the potential of A.I. while safeguarding individual rights, promoting transparency, and emphasizing inclusivity.

Throughout this work, we explore the risks and rewards of A.I. adoption, from the potential of financial and social manipulation to the necessity of environmental stewardship and ethical oversight. Importantly, we stress the need for ethical human leadership in guiding A.I. policies—leaders who prioritize the welfare of society over mere technological advancement. This framework is not just a set of rules; it is a call to action, a reminder of our shared responsibility to ensure that A.I. empowers rather than exploits, uplifts rather than undermines.

As we move forward, it is my hope that this framework will serve as a foundation for informed decision-making, fostering a future where A.I. is a force for good. Let us use this moment to establish policies that protect human dignity, promote justice, and champion innovation in harmony with ethical principles. Together, we have the power to shape a future where technology and humanity coexist in mutual respect, creating a world that reflects the best of what we aspire to achieve.

Thank you for your commitment to this cause. May this document guide your efforts and inspire a vision for a fairer, safer, and more ethical future.

Tony Vortex
contact@ethicalaigov.com
https://ethicalaigov.com

OVERVIEW OF THE PURPOSE OF THE FRAMEWORK

As artificial intelligence (A.I.) transforms industries, communities, and individuals' daily lives, society finds itself in uncharted territory. A.I. holds the promise to solve complex global challenges, streamline processes, and improve human well-being. However, with great power comes great responsibility. Without carefully considered policies and ethical oversight, A.I. could exacerbate social inequalities, erode personal autonomy, disrupt economies, and create new avenues for exploitation and control.

This framework has been created to guide policymakers, corporations, community leaders, and individuals in developing and deploying A.I. responsibly. It seeks to establish a balanced approach to A.I. governance that prioritizes human-centered values, safeguards public welfare, and promotes ethical technological advancement. By laying out both the opportunities and risks associated with A.I., this document serves as a roadmap to help stakeholders harness the potential of A.I. while proactively addressing its most pressing challenges.

Scope and Objectives

The scope of this framework is comprehensive, covering all major aspects of A.I. impact, from societal and economic transformations to environmental and ethical considerations. The objectives are to:

1. **Promote Ethical A.I. Development and Governance**: Advocate for policies that place ethics and human rights at the forefront of A.I. development.

2. **Establish Clear Guidelines for A.I. Applications**: Outline standards and protocols to guide the use of A.I. in various sectors, from healthcare and security to finance and social media.

3. **Protect Individual Rights and Societal Well-being**: Address privacy, autonomy, and security concerns to ensure that A.I. serves humanity without compromising fundamental freedoms.

4. **Incorporate Diverse Voices and Perspectives**: Ensure inclusivity in A.I. governance by involving ethical human advocates and communities traditionally marginalized by technological advancement.

5. **Future-Proof A.I. Governance**: Prepare for unforeseen advancements and challenges by creating adaptable policies and building resilience in governance structures.

Current State of A.I. and Projected Growth

A.I. technology has evolved rapidly over the past decade, with applications now influencing everything from healthcare diagnostics and financial services to content curation on social media. Today's A.I. systems are capable of performing complex tasks, analyzing vast amounts of data, and making decisions that were once exclusively within the human domain. Yet, as A.I. capabilities continue to expand, so too do the ethical and societal questions surrounding its use.

With A.I. projected to continue its exponential growth, there is an urgent need for governance structures that can keep pace with these advancements. The current regulatory landscape is fragmented, with some countries adopting stringent data protection laws, while others lack comprehensive A.I. oversight. This disparity creates a global challenge, as A.I. is a borderless technology that requires cohesive, collaborative governance to prevent misuse and ensure ethical deployment.

Historical Context: Lessons from Recent Developments and Social Impact

The COVID-19 pandemic underscored the power of large corporations and digital platforms to influence public discourse, economics, and even social behavior. During lockdowns, major technology companies took on outsized roles, and some used this influence in ways that raised concerns over ethics and fairness. Cases of financial and social "terrorism" emerged, with accusations that companies leveraged A.I. and big data to exert control over markets, suppress competition, and manipulate public opinion. These developments served as a wake-up call, highlighting the need for stringent A.I. governance to prevent corporate overreach and protect public interests.

Moreover, issues like the "Twitter Files" controversy and PayPal's proposed misinformation fine have shown how A.I.-enabled platforms can be weaponized to restrict information, limit discourse, and control user behavior. These events illustrate the potential for A.I. to be used in ways that are antithetical to democratic values, underscoring the need for a governance framework that places ethical human leadership at its core.

The Role of Ethical Human Leadership in A.I. Governance

In response to these concerns, this framework introduces a crucial component: the need for ethical human advocates in A.I. governance roles. By including individuals with proven track records in social advocacy, human rights, and public welfare, this framework emphasizes the importance of ethical leadership in overseeing A.I. development. Leaders with a commitment to humanity-first values are better positioned to ensure that A.I. serves as a tool for social good rather than corporate or political manipulation. Their involvement brings a moral compass to A.I. governance, ensuring that technology development aligns with human-centric principles of fairness, empathy, and justice.

These ethical leaders act as safeguards, preventing the exploitation of A.I. for exclusionary or discriminatory purposes. Their roles span establishing ethical standards, monitoring compliance, and educating the public—bridging the gap between technological advancement and the human values that ground our societies.

The Need for a Balanced Approach to A.I. Development and Governance

The potential of A.I. to revolutionize society is matched only by its potential risks. This framework advocates for a balanced approach to A.I. development—one that embraces innovation while protecting individual rights and societal values. Each chapter within this framework is designed to address a specific aspect of A.I. governance, providing in-depth analysis, policy recommendations, and practical guidelines.

The framework includes measures to mitigate risks such as financial and social terrorism, where corporations leverage A.I. to manipulate markets or suppress dissent. It addresses the need for privacy and autonomy in an era of pervasive surveillance and data collection. Additionally, it highlights the environmental impacts of A.I., advocating for sustainable practices in technology development.

This balanced approach also requires a focus on inclusivity, ensuring that A.I. benefits all members of society, not just those with economic or political power. By emphasizing inclusivity, this framework acknowledges the diversity of human experiences and the importance of protecting vulnerable populations from potential harms associated with A.I.

Vision for a Harmonious Coexistence of Humans and Technology

This framework envisions a future where A.I. and humanity coexist in harmony, enhancing each other's strengths. In this vision, A.I. serves as an extension of human ingenuity, empowering individuals, fostering societal progress, and helping address some of the world's most pressing challenges. To achieve this future, we must commit to A.I. governance that values human rights, promotes ethical integrity, and prioritizes public welfare.

By grounding A.I. in human-first principles, we can create a world where technology not only enhances productivity but also uplifts humanity. The path forward requires collaboration across all sectors, involving governments, corporations, and individuals in a shared mission to build an A.I.-enabled society that is safe, just, and inclusive.

In embarking on this journey, this framework calls on all stakeholders to engage with A.I. responsibly, thoughtfully, and with unwavering dedication to the common good. Together, we can shape a future where A.I. becomes a force for positive transformation, driven by values that respect and celebrate our shared humanity.

References

Please note, the following references provide general insights into topics like artificial intelligence, ethics, and governance but may not directly address synthetic human regulation as described. These references are useful starting points for understanding the broader context of A.I. governance.

1.   Bostrom, N., & Yudkowsky, E. (2012).
     The Ethics of Artificial Intelligence. Oxford University Press.

2.   Kurzweil, R. (2005). *The Singularity Is Near: When Humans Transcend Biology*. Penguin.

3.   Menzel, P., & D'Aluisio, F. (2002).
     Robo sapiens: Evolution of a New Species. Cambridge, MA: MIT Press.

4.   Russell, S., & Norvig, P. (2009). *Artificial Intelligence: A Modern Approach*. Prentice Hall.

5.   Cooper, J. (2010).
     Humanity's End: Why We Should Reject Radical Enhancement. MIT Press.

CHAPTER 1:
ETHICAL CONSIDERATIONS
IN A.I. DEVELOPMENT

1.1 Introduction to Ethical Considerations in A.I. Development

As artificial intelligence (A.I.) technology advances, the ethical implications of its development have become increasingly significant. The rapid integration of A.I. into critical sectors, including healthcare, finance, and public safety, necessitates a thorough understanding of the ethical considerations that must guide its deployment. This chapter explores the ethical dimensions of A.I. development, including concerns about bias, accountability, transparency, and the need for human oversight.

The goal of this chapter is to establish a foundation for ethical A.I. governance, ensuring that A.I. systems are designed and implemented in ways that respect human rights, promote fairness, and enhance societal well-being.

1.2 Addressing Bias and Fairness in A.I. Systems

One of the most pressing ethical issues in A.I. development is algorithmic bias. A.I. systems often rely on large datasets that can unintentionally reflect existing societal biases. If these biases are not identified and mitigated, A.I. systems can produce discriminatory outcomes, particularly in high-stakes areas such as hiring, lending, and law enforcement.

To address bias, developers must prioritize diverse and representative data, implement fairness-checking algorithms, and conduct regular audits. Ethical A.I. development requires a commitment to fairness, where the potential for discrimination is minimized, and equality is actively promoted.

1.3 Transparency and Accountability in A.I. Decision-Making

Transparency in A.I. systems is crucial to maintaining public trust and ensuring accountability. Users and stakeholders have a right to understand how A.I. systems make decisions, particularly in scenarios where these systems impact human lives. Lack of transparency can lead to "black box" systems, where A.I. decisions are opaque and difficult to interpret.

Establishing accountability mechanisms is essential in A.I. development. Developers and organizations should be responsible for the actions of their A.I. systems, with clear protocols for addressing errors or harmful outcomes. By embedding accountability into the design and deployment of A.I., organizations can ensure that systems operate within ethical boundaries and that appropriate recourse is available when issues arise.

1.4 Privacy and Data Security Concerns

A.I. systems rely on large amounts of personal and sensitive data to function effectively. While data collection enables valuable insights and personalized services, it also raises concerns about privacy and data security. Without strict data governance frameworks, individuals may lose control over their personal information, leading to potential misuse or unauthorized access.

Ethical A.I. development must prioritize data protection, ensuring that A.I. systems are compliant with data privacy regulations and that individuals have control over their data. Privacy-by-design principles should be integrated into A.I. development, making privacy and data security foundational aspects of system architecture rather than afterthoughts.

1.5 Human Oversight and the Importance of Control

While A.I. systems can operate autonomously, human oversight remains essential in ensuring that they function ethically and responsibly. A.I. should be viewed as a tool that assists human decision-making, not as a replacement for human judgment. Critical decisions, particularly in areas such as healthcare, law enforcement, and finance, should always involve human review to prevent unintended harm and maintain ethical standards.

Establishing clear protocols for human oversight, including escalation procedures and intervention points, can prevent A.I. systems from acting in ways that contradict ethical principles. By maintaining human control over A.I., organizations can ensure that technology serves humanity's interests.

1.6 Ethical Design Principles for A.I. Development

Ethical A.I. development requires adherence to core design principles that prioritize human values and rights. These principles include fairness, transparency, accountability, and privacy, all of which must be embedded into the development lifecycle from the beginning. Developers should follow established ethical frameworks, such as the IEEE's Ethically Aligned Design or the European Commission's Ethics Guidelines for Trustworthy A.I., to guide responsible practices.

By committing to ethical design, organizations can foster A.I. systems that respect individual autonomy, promote social equity, and align with broader societal values. Ethical design is not only about preventing harm but also about proactively creating A.I. that contributes positively to society.

Conclusion of Chapter 1

The ethical considerations in A.I. development extend beyond technical challenges to fundamental questions of human rights, social equity, and individual autonomy. As A.I. becomes increasingly integrated into daily life, developers and organizations must commit to ethical principles that safeguard against discrimination, protect privacy, and maintain transparency. By embedding these values into A.I. development from the outset, we can build systems that serve humanity responsibly, aligning with our highest ethical standards.

References

1. Binns, R. (2018). *Fairness in Machine Learning: Lessons from Political Philosophy*. Proceedings of the 2018 Conference on Fairness, Accountability, and Transparency.

2. Floridi, L., & Cowls, J. (2019). *A Unified Framework of Five Principles for AI in Society*. Harvard Data Science Review.

3. IEEE. (2019). *Ethically Aligned Design: A Vision for Prioritizing Human Well-being with Autonomous and Intelligent Systems*. IEEE Standards Association.

4. Mittelstadt, B., Allo, P., Taddeo, M., Wachter, S., & Floridi, L. (2016). *The Ethics of Algorithms: Mapping the Debate*. Big Data & Society.

5. Wachter, S., Mittelstadt, B., & Floridi, L. (2017). *Why a Right to Explanation of Automated Decision-Making Does Not Exist in the General Data Protection Regulation*. International Data Privacy Law, 7(2), 76-99.

6. Zarsky, T. Z. (2016). *The Trouble with Algorithmic Decisions: An Analytic Road Map to Examine Efficiency and Fairness in Automated and Opaque Decision Making*. Science, Technology, & Human Values, 41(1), 118-132.

7. European Commission. (2019). *Ethics Guidelines for Trustworthy AI*. Independent High-Level Expert Group on Artificial Intelligence.

CHAPTER 2:
SOCIETAL IMPACTS OF A.I.
AND AUTOMATION

2.1 Employment and Economic Displacement

One of the most immediate societal impacts of advancing A.I. and automation technology is the potential disruption of traditional employment sectors. As companies adopt A.I. for tasks once performed by humans, economic displacement may occur, affecting a wide range of industries from manufacturing to healthcare. This shift will require targeted policies and adaptation strategies to minimize economic disruptions.

- **Job Displacement in Routine Roles**: A.I. systems excel at handling repetitive tasks with speed and precision, leading to automation of jobs in sectors such as manufacturing, warehousing, customer service, and data processing. As these roles are increasingly automated, a significant portion of the workforce may find itself displaced, particularly in entry-level and low-skilled positions.

- **The Need for Reskilling Programs**: To mitigate the negative impact on employment, governments and corporations must prioritize reskilling and upskilling programs that prepare workers for roles where human skills, such as creativity, critical thinking, and emotional intelligence, remain essential. Workforce adaptation policies will play a critical role in preventing economic stagnation and supporting career transitions.

- **Economic Inequality and Accessibility**: A.I. adoption may disproportionately benefit large corporations with the resources to invest in advanced technology, widening the gap between big businesses and small enterprises. Policies focused on equitable access to A.I. tools for small businesses and local enterprises can help reduce economic inequality and support a more inclusive economy.

2.2 Public Perception and Trust in A.I. Systems

The integration of A.I. into public life depends heavily on public perception and trust. As A.I. systems become embedded in essential services, such as healthcare, education, and finance, society must grapple with the implications of re-lying on autonomous systems for decision-making. Building and maintaining public trust is essential for successful A.I. adoption.

- **Transparency and Explainability**: A key factor in fostering public trust is ensuring that A.I. systems operate transparently. Individuals interacting with A.I. systems need clear, understandable explanations of how decisions are made, especially in high-stakes situations like healthcare diagnoses, loan approvals, and legal rulings. Transparent A.I. can build trust and encourage acceptance of technology.

- **Bias and Fairness**: Public trust can be compromised when A.I. systems exhibit biases in decision-making. When algorithms inadvertently reinforce social biases—such as those based on race, gender, or socioeconomic status—they can perpetuate inequality and erode public confidence. Regular auditing, diverse training data, and bias mitigation strategies are crucial to ensuring fairness in A.I. systems.

- **Education and Awareness Initiatives**: To address concerns and prevent fearmongering, public education initiatives on A.I. capabilities and limitations are essential. Misunderstandings and myths surrounding A.I. often lead to exaggerated fears; educating the public on the practical applications and constraints of A.I. can support a balanced perspective.

2.3 Social and Cultural Shifts in a Technology-Driven Society

The rapid rise of A.I. and automation is poised to create significant shifts in cultural norms, societal values, and human behavior. As A.I. systems become integral to daily life, they will inevitably shape social structures and impact human interaction, requiring careful consideration of potential psychological and cultural effects.

- **Changing Social Norms Around Interaction**: As A.I.-powered virtual assistants and customer service bots become commonplace, people may begin to normalize interaction with non-human entities. This shift may alter communication styles, patience levels, and expectations in human relationships, as society adjusts to interacting with systems that simulate human-like responses.

- **Reduced Human-to-Human Interaction**: Increased reliance on A.I. for transactional interactions, such as shopping, customer support, and even some medical consultations, could lead to a reduction in face-to-face human interactions. While this convenience offers time savings, it may affect social cohesion, reduce opportunities for human connection, and even impact mental health over time.

- **Impacts on Human Identity and Autonomy**: The integration of A.I. into personal and professional spaces raises questions about autonomy and self-identity. When individuals rely on A.I. for decision-making, such as financial management or career advice, it may affect their sense of control and independence. Additionally, people may begin to question what it means to be uniquely "human" as A.I. systems mimic more complex behaviors.

2.4 Ethical Implications of A.I.-Driven Decision-Making

As A.I. systems play a larger role in decision-making processes across sectors, ethical concerns arise around their use, especially in sensitive areas like healthcare, criminal justice, and finance. The ethical challenges associated with A.I.-driven decisions highlight the need for accountability and oversight.

- **Healthcare and Autonomous Diagnoses**: In healthcare, A.I. is increasingly used for diagnostics, treatment recommendations, and even surgeries. While A.I. can enhance accuracy and efficiency, ethical concerns emerge regarding patient autonomy, consent, and potential overreliance on automated systems. Clear guidelines and human oversight must be maintained to ensure patient safety.

- **Criminal Justice and Predictive Policing**: Predictive algorithms in criminal justice raise ethical questions about fairness, accountability, and transparency. A.I.-based predictive policing has been criticized for perpetuating biases, as these systems often reflect historical policing data that may be inherently biased. Strict ethical standards and human oversight are needed to prevent unjust outcomes.

- **Financial Decisions and Algorithmic Bias**: A.I. is now used for credit scoring, loan approvals, and financial recommendations. When financial algorithms incorporate biases from historical data, they can inadvertently disadvantage certain demographic groups. Auditing and transparent scoring criteria can help to mitigate these issues and promote fairness.

2.5 Emotional Influence and Behavioral Manipulation Risks

A.I. technologies, particularly those with emotional intelligence capabilities, bring forth ethical concerns related to their influence on human emotions and behavior. From recommendation algorithms to social media bots, these technologies can subtly shape opinions, purchasing behaviors, and social interactions, presenting risks to personal autonomy.

- **Algorithmic Persuasion in Media and Advertising**: Social media platforms and online retailers leverage A.I. to deliver targeted advertisements and content recommendations, tailoring experiences to individual users. While this can improve user engagement, it raises ethical questions about manipulation and informed consent. Individuals may be unaware of how these systems shape their preferences and opinions.

- **Mental Health Impacts of A.I.-Curated Content**: Research shows that A.I.-driven content curation can influence mood, self-esteem, and overall mental health. For example, algorithms that prioritize sensationalist content to boost engagement can contribute to anxiety, stress, and misinformation. Ethical content curation standards and transparency in recommendation algorithms are essential for protecting users' well-being.

- **Vulnerability of Children and Adolescents**: Younger users are especially vulnerable to the influence of A.I.-curated content and recommendations. Social media algorithms can shape self-image, behavior, and mental health. Implementing safeguards for children, such as parental controls and content restrictions, is critical to prevent adverse effects on developing minds.

Conclusion of Chapter 2

The societal impacts of A.I. and automation extend beyond simple technological advances; they touch upon fundamental aspects of human life, from economic security and social structures to identity and ethical considerations. While A.I. systems offer efficiency and convenience, they must be carefully managed to prevent unintended consequences that could disrupt social cohesion or infringe upon individual rights.

To achieve this balance, policymakers must address the ethical and social dimensions of A.I. integration, promoting transparency, accountability, and public awareness. By taking a proactive approach to these challenges, society can harness the benefits of A.I. while safeguarding against risks that could undermine trust, security, and societal well-being.

References

1.   Brynjolfsson, E., & McAfee, A. (2014). *The Second Machine Age: Work, Progress, and Prosperity in a Time of Brilliant Technologies*. W. W. Norton & Company.

2.   Eubanks, V. (2018). *Automating Inequality: How High-Tech Tools Profile, Police, and Punish the Poor*. St. Martin's Press.

3.   Noble, S. U. (2018). *Algorithms of Oppression: How Search Engines Reinforce Racism*. NYU Press.

4.   O'Neil, C. (2016). *Weapons of Math Destruction: How Big Data Increases Inequality and Threatens Democracy*. Crown.

5.   Crawford, K. (2021). *Atlas of AI: Power, Politics, and the Planetary Costs of Artificial Intelligence*. Yale University Press.

CHAPTER 3: TECHNOLOGICAL ADVANCEMENTS AND CAPABILITIES

3.1 Overview of Current and Emerging A.I. Technologies

Advancements in artificial intelligence (A.I.) have led to a surge of innovations across industries, pushing boundaries in machine learning, natural language processing, robotics, and computer vision. These technologies are not only transforming individual sectors but are also converging to create increasingly autonomous systems capable of performing complex tasks with minimal human intervention.

- **Machine Learning and Predictive Analytics**: Machine learning models now power predictive analytics across various domains, from healthcare diagnostics to financial forecasting. By analyzing historical data, these models can predict future outcomes with impressive accuracy, driving better decision-making and enabling preventive measures in fields like medicine, finance, and agriculture.

- **Natural Language Processing (NLP)**: NLP has advanced significantly, allowing A.I. to understand, process, and respond to human language. This capability underpins virtual assistants, customer service chatbots, and content generation tools. As A.I. becomes more adept at understanding context and nuance, NLP applications are expanding to include sentiment analysis, real-time translation, and automated summarization in business and media.

- **Computer Vision**: A.I.-powered computer vision enables machines to interpret visual data, opening applications in security, quality control, and autonomous vehicles. For example, facial recognition is increasingly used in secure access systems, while image recognition assists in medical imaging diagnostics. Computer vision is also crucial in areas like manufacturing, where it can identify defects and improve quality assurance.

- **Robotic Process Automation (RPA)**: RPA is revolutionizing business operations by automating repetitive, rule-based tasks in fields like data entry, payroll processing, and customer support. By handling mundane tasks, RPA allows human workers to focus on higher-level functions, increasing overall productivity and efficiency.

- **Generative A.I.**: Generative A.I. models, such as those behind deepfakes and content creation tools, can generate text, images, audio, and even video based on input prompts. While this has promising applications in creative industries, it also raises ethical concerns about misinformation and intellectual property rights, making regulation essential.

3.2 Autonomous Systems in Industrial and Commercial Sectors

Autonomous systems, driven by advancements in A.I., are rapidly entering industrial and commercial settings, where they offer efficiency, accuracy, and scalability. These systems are enhancing productivity across sectors such as manufacturing, logistics, healthcare, and agriculture.

- **Manufacturing and Assembly**: Autonomous robots are increasingly being used on factory floors for assembly, welding, painting, and quality control. These robots work alongside human operators or function independently in settings where precision and consistency are critical, helping to reduce errors and waste.

- **Warehousing and Logistics**: Autonomous vehicles and robotic systems are transforming warehousing and logistics. From self-driving forklifts to automated picking and sorting systems, A.I. is optimizing inventory management, order fulfillment, and last-mile delivery. E-commerce giants and logistics companies are adopting these technologies to meet rising consumer demand efficiently.

- **Agriculture and Precision Farming**: A.I.-driven automation is also reshaping agriculture, with autonomous tractors, drones, and soil monitoring systems becoming more common. Precision farming allows for optimized planting, fertilization, and irrigation, which increases crop yields and reduces resource usage. These advancements contribute to sustainable agriculture practices that are both economically and environmentally beneficial.

- **Healthcare and Surgery**: Autonomous systems are assisting in areas such as robotic surgery, where precision is paramount. Robotic-assisted surgeries, powered by A.I., provide enhanced accuracy in minimally invasive procedures, reducing patient recovery time. In addition, autonomous A.I. systems support administrative tasks, diagnostics, and even patient interaction, increasing efficiency in healthcare delivery.

3.3 A.I.-Enhanced Decision-Making in High-Stakes Environments

A.I. systems are increasingly employed to assist with decision-making in high-stakes environments where rapid analysis and accuracy are critical. While human oversight remains essential, A.I.-enhanced decision-making is proving valuable in areas such as finance, law enforcement, and disaster response.

- **Financial Services and Risk Management**: In finance, A.I. algorithms analyze vast datasets to assess risks, detect fraudulent transactions, and make real-time trading decisions. These systems support human analysts by identifying patterns that are otherwise difficult to detect. For example, A.I.-driven risk management systems can monitor markets and manage portfolio risks, assisting banks and financial institutions in protecting investments.

- **Law Enforcement and Public Safety**: Predictive policing algorithms, though controversial, are being tested by some law enforcement agencies to identify patterns in criminal activity and allocate resources accordingly. These systems analyze crime data to identify high-risk areas and predict where incidents may occur, though they raise ethical concerns about privacy and bias.

- **Emergency Response and Disaster Management**: A.I. is proving beneficial in emergency response by analyzing weather patterns, infrastructure data, and social media feeds to predict and respond to natural disasters. For example, A.I. systems can predict the paths of hurricanes or wildfires, enabling better resource allocation and evacuation planning. In post-disaster situations, A.I.-powered drones and mapping systems aid in search-and-rescue operations, identifying areas with the highest need.

3.4 Emerging Applications in Personal and Public Spaces

A.I.-driven systems are increasingly found in both personal and public spaces, providing convenience and enhancing quality of life. These applications range from smart home devices to urban infrastructure, bringing both opportunities and challenges for societal integration.

- **Smart Homes and Personal Assistants**: Personal A.I. assistants, such as Amazon's Alexa, Google Assistant, and Apple's Siri, are now commonplace in households worldwide. These devices perform various tasks, from controlling home appliances to providing information and reminders, integrating seamlessly into daily life. However, concerns about data privacy, security, and dependence on such devices remain prevalent.

- **Smart Cities and Urban Planning**: A.I. is transforming urban infrastructure through applications like traffic management, energy optimization, and predictive maintenance. Smart traffic systems, for instance, use real-time data to optimize traffic flow and reduce congestion. In addition, A.I.-powered sensors can monitor infrastructure health, alerting authorities to potential issues before they become critical.

- **Education and Personalized Learning**: A.I. is also making strides in education, where personalized learning platforms adapt to each student's learning pace and style. By analyzing performance data, these systems can provide tailored resources, identify areas for improvement, and adjust lesson plans to maximize student engagement. While promising, these applications must be carefully managed to prevent over-reliance and ensure that human educators remain central to the learning experience.

- **Public Health and Pandemic Response**: A.I. systems played a pivotal role during recent global health crises, such as COVID-19, by aiding in diagnostics, contact tracing, and vaccine development. Machine learning algorithms helped identify viral patterns, while A.I.-driven data analysis assisted in modeling the spread of the virus. These applications highlight A.I.'s potential to support public health, though they also emphasize the importance of data security and ethical data use.

3.5 Limitations and Ethical Considerations in Advanced A.I. Development

As A.I. technology progresses, ethical considerations and technical limitations remain key challenges. While autonomous and intelligent systems offer unprecedented benefits, they also introduce risks that require careful oversight and regulation.

- **Limitations of A.I. Systems**: Despite significant advancements, A.I. is still limited in areas such as common-sense reasoning, contextual understanding, and genuine creativity. These limitations underscore the need for human oversight, particularly in decision-making roles where nuanced judgment is critical.

- **Ethical Implications of Surveillance and Privacy**: A.I. systems often rely on vast datasets, raising concerns about privacy and surveillance. Facial recognition, behavior tracking, and data collection are powerful tools but risk infringing on personal privacy and civil liberties if misused. Regulatory guidelines for responsible data use and transparent data collection practices are essential.

- **Bias and Fairness in Algorithms**: A.I. algorithms can unintentionally perpetuate or amplify biases present in training data. This risk is particularly problematic in high-stakes environments, such as law enforcement or hiring, where biased decisions can lead to significant harm. Implementing regular audits, using diverse datasets, and establishing standards for fairness can help mitigate these biases.

- **Human Dependency on A.I.**: As A.I. systems increasingly handle daily tasks and assist with decision-making, there is a growing risk of dependency. People may become overly reliant on A.I. for personal or professional tasks, potentially leading to a decline in certain skills or critical thinking abilities. Encouraging balanced A.I. use and promoting human agency are essential for preventing over-reliance.

Conclusion of Chapter 3

The technological advancements in A.I. and automation are transforming industries, enhancing efficiency, and improving quality of life. However, these advancements come with limitations and ethical considerations that require careful regulation and human oversight. As A.I. systems become increasingly autonomous, it is essential to maintain a balanced approach, one that leverages A.I.'s capabilities without sacrificing ethical standards or human rights.

Policymakers must address the challenges associated with these advancements to ensure that A.I. technology develops responsibly and aligns with societal values. By setting clear standards for transparency, accountability, and ethical conduct, society can harness A.I. for the collective good while mitigating potential risks.

References

1. Russell, S., & Norvig, P. (2009). *Artificial Intelligence: A Modern Approach*. Prentice Hall.

2. Brynjolfsson, E., & McAfee, A. (2014). *The Second Machine Age: Work, Progress, and Prosperity in a Time of Brilliant Technologies*. W. W. Norton & Company.

3. Crawford, K. (2021). *Atlas of AI: Power, Politics, and the Planetary Costs of Artificial Intelligence*. Yale University Press.

4. O'Neil, C. (2016). *Weapons of Math Destruction: How Big Data Increases Inequality and Threatens Democracy*. Crown.

5. Floridi, L. (2019). *The Ethics of Artificial Intelligence: Moral Perspectives on Human and Non-Human Agents*. Cambridge University Press.

CHAPTER 4: SECURITY RISKS AND THREAT MANAGEMENT

4.1 Espionage and Surveillance Concerns

A.I. and advanced automation technologies have enhanced surveillance capabilities, allowing for unprecedented levels of data collection, analysis, and monitoring. While these tools can be valuable for law enforcement and national security, they also pose significant privacy and espionage risks if misused by malicious actors, corporations, or even state agencies.

- **A.I.-Driven Surveillance Systems**: A.I. has revolutionized surveillance through technologies like facial recognition, behavior analysis, and real-time data monitoring. These systems are deployed in various public and private spaces, from city surveillance to secure corporate campuses. However, when used without proper oversight, they can infringe on personal privacy and create environments of constant monitoring that may affect citizens' behavior and freedoms.

- **Corporate and Industrial Espionage**: Corporations increasingly employ A.I. to monitor market trends, competitor activities, and proprietary data. In this context, A.I.-powered espionage risks escalate as automated systems can swiftly extract valuable insights from vast datasets, including intellectual property and trade secrets. Safeguards must be established to prevent corporate espionage facilitated by A.I.

- **Cybersecurity Protocols for A.I. Surveillance**: As A.I.-powered surveillance becomes more ubiquitous, cybersecurity measures must evolve to protect the data collected and ensure these systems aren't exploited. Ensuring end-to-end encryption, secure data storage, and strict access controls can mitigate risks of unauthorized access to surveillance data.

4.2 Sabotage and Physical Security Threats

A.I.-enabled systems are increasingly used in critical infrastructure and manufacturing sectors, making them potential targets for sabotage. Malicious actors could exploit vulnerabilities in autonomous systems to disrupt essential services, cause physical damage, or interfere with industrial processes. These risks necessitate rigorous security protocols for A.I. deployment in sensitive environments.

- **Vulnerabilities in Autonomous Systems**: A.I.-controlled machines and robots, from industrial machinery to autonomous vehicles, may be susceptible to hacking and remote control attacks. A compromised autonomous vehicle, for example, could be directed to cause traffic disruptions or accidents. Similarly, attacks on automated industrial equipment could halt production or damage facilities.

- **Infrastructure Targeting**: Autonomous systems controlling utilities (e.g., power grids, water treatment facilities) are highly vulnerable to cyber-attacks. Such sabotage could have cascading effects on public safety, health, and national security. Strict cybersecurity measures and intrusion detection systems are vital for protecting critical infrastructure.

- **Preventive Security Measures**: Implementing multi-layered security protocols, including biometric authentication, firewalls, and real-time monitoring, can help reduce the risk of sabotage. Additionally, regular penetration testing and vulnerability assessments of autonomous systems can identify and address potential weaknesses before they are exploited.

4.3 Manipulation of Public Opinion and Social Engineering

A.I. systems are capable of influencing public opinion through targeted manipulation and social engineering techniques, often deployed in political campaigns, advertising, and misinformation. While A.I.-enabled social media bots and recommendation algorithms can enhance engagement, they also risk manipulating user perceptions and decision-making.

- **Social Media Bots and Fake News Propagation**: A.I. can create, amplify, and spread misinformation through bots and fake social media accounts, misleading the public on critical issues. These bots can generate thousands of automated posts, shaping public opinion by creating the illusion of consensus or driving divisive narratives. Preventing the misuse of A.I. for misinformation campaigns is crucial to protect democratic processes and maintain trust in information sources.

- **Microtargeting and Psychological Manipulation**: A.I. algorithms analyze user data to deliver highly personalized content, advertisements, or political messages. While this level of targeting enhances relevance, it also raises concerns about psychological manipulation. Individuals may be unwittingly influenced to act in ways that serve the agenda of advertisers, corporations, or political entities.

- **Countermeasures Against Manipulation**: Transparency in content origins, digital literacy education, and algorithmic accountability are key defenses against social engineering through A.I. Regulations requiring platforms to disclose when content is algorithmically generated or microtargeted can help users identify potential manipulation and make informed decisions.

4.4 Autonomous Warfare and Lethal Capabilities

A.I. has the potential to transform warfare, with autonomous systems being increasingly used in defense for tasks ranging from surveillance and reconnaissance to offensive operations. While these systems offer strategic advantages, their deployment raises significant ethical and security risks.

- **A.I.-Enabled Autonomous Weapons**: Autonomous weapons systems (AWS), such as drones, can identify and engage targets without human intervention. While they provide operational advantages, the risk of errors or unintended escalation is substantial. Without adequate safeguards, A.I.-enabled weapons could cause accidental harm or provoke international tensions.

- **Decision-Making in Combat Scenarios**: Allowing A.I. to make life-or-death decisions in combat situations raises serious ethical questions. Autonomous systems lack the moral judgment that human soldiers possess, and A.I.-driven decisions may prioritize efficiency over humanitarian considerations. Ensuring human oversight in decision-making can mitigate the risks of unintended casualties or violations of international law.

- **International Regulations and Agreements**: Countries must work together to establish international regulations governing the use of A.I. in warfare. Clear guidelines around the development and deployment of autonomous weapons, along with accountability mechanisms, can help prevent the escalation of autonomous warfare and reduce the risk of misuse.

4.5 Mechanisms for Detection and Identification of Synthetic Manipulations

As A.I. becomes more sophisticated, it will be increasingly challenging to detect when systems are used to manipulate or deceive. Effective detection mechanisms and protocols for identifying synthetic activity, whether digital or physical, are essential for security.

- **Detection of Deepfakes and Synthetic Media**: Deepfake technology, which uses A.I. to create hyper-realistic synthetic media, poses risks of impersonation and misinformation. For example, deepfake videos could be used to impersonate political leaders, spreading false statements and creating confusion. Development of detection tools that can identify synthetic media is essential to combat these threats.

- **Monitoring A.I.-Driven Misinformation Campaigns**: Governments and organizations can monitor social media and other digital platforms to identify and counteract A.I.-driven misinformation campaigns. Utilizing machine learning models to flag unusual content patterns or automated activity can help detect synthetic manipulation at an early stage.

- **Physical Detection in Sensitive Environments**: As A.I.-enabled systems become integrated into physical environments, detection mechanisms are necessary to ensure they behave as expected. This is particularly important in secure facilities or government buildings, where A.I.-powered devices or robots may require regular verification to prevent unauthorized access or sabotage.

Conclusion of Chapter 4

The security risks associated with A.I. technology are complex and multifaceted, impacting areas from privacy and manipulation to physical security and warfare. As A.I. capabilities grow, so too do the potential threats posed by malicious actors seeking to exploit these technologies. The risks covered in this chapter underscore the importance of robust security frameworks, continuous monitoring, and proactive measures to protect society from A.I.-enabled threats.

To address these risks effectively, governments, corporations, and technology developers must collaborate on creating safeguards and developing ethical guidelines. By establishing clear standards, accountability measures, and international cooperation, we can work to mitigate the dangers of A.I.-powered manipulation, sabotage, and warfare while benefiting from the technology's transformative potential.

References

1. Singer, P. W., & Brooking, E. T. (2018). *LikeWar: The Weaponization of Social Media.* Houghton Mifflin Harcourt.

2. Scharre, P. (2018). *Army of None: Autonomous Weapons and the Future of War.* W. W. Norton & Company.

3. O'Neil, C. (2016). *Weapons of Math Destruction: How Big Data Increases Inequality and Threatens Democracy.* Crown.

4. Crawford, K. (2021). *Atlas of AI: Power, Politics, and the Planetary Costs of Artificial Intelligence.* Yale University Press.

5. Floridi, L. (2019). *The Ethics of Artificial Intelligence: Moral Perspectives on Human and Non-Human Agents.* Cambridge University Press.

CHAPTER 5:
RISK OF FINANCIAL
AND SOCIAL TERRORISM

5.1 Financial Terrorism: Corporate Power, Big Data, and Economic Manipulation

The COVID-19 pandemic created an unprecedented global crisis, and during the lockdowns from 2020 to 2022, large corporations, particularly in the technology and e-commerce sectors, capitalized on this disruption. Leveraging their vast resources, these companies employed big data and A.I. tools to maintain and even expand their dominance. Critics argue that these corporations engaged in what can be described as a form of "financial terrorism"—using their market power to suppress competition, stifle individual economic freedom, and manipulate economic conditions to their advantage.

- **Monopolistic Practices and Economic Suppression**: As small businesses struggled under lockdown restrictions, large corporations in sectors like e-commerce, cloud computing, and digital advertising saw exponential growth. Companies like Amazon, Google, and Facebook became lifelines for consumers confined to their homes. However, by controlling essential online platforms, these corporations dictated the terms of participation, often sidelining smaller competitors unable to compete with their logistics, data analytics, and customer reach.

    o   Amazon's e-commerce model, for instance, squeezed out countless small retailers unable to operate or keep up with Amazon's delivery speed and market reach. The pandemic allowed Amazon to expand its influence, essentially creating an ecosystem where small businesses could only survive by listing products on Amazon's platform—where they were subject to high fees, loss of direct customer interaction, and intense competition against Amazon's own products (Soper, 2020).

- **Data and A.I.-Driven Competitive Advantage**: Large tech companies used A.I. algorithms and big data analytics to monitor consumer behavior, anticipate market trends, and target advertisements with pinpoint precision. By collecting and analyzing data on a massive scale, these corporations could adjust pricing strategies, manage inventory, and capture consumer loyalty more effectively than traditional retailers, who lacked access to such granular data. This data-driven control over the marketplace has allowed tech giants to dictate the trajectory of entire sectors and solidify their positions of power.

- o For example, Facebook and Google controlled digital advertising markets during the lockdowns, leaving few advertising options for small businesses who needed online exposure but could not afford to match large corporations' budgets. This effectively limited the visibility of smaller players, reinforcing the dominance of established market leaders (Levy, 2021).

5.2 Social Terrorism: Manipulation of Social Platforms and Public Discourse

The COVID-19 pandemic also saw the rise of what some describe as "social terrorism"—where major technology companies used their social influence, enhanced by A.I.-driven content moderation and recommendation algorithms, to control and shape public discourse. By leveraging their dominance over social media and digital communication platforms, these companies played a significant role in curating the information people saw, influencing opinions, and often silencing dissenting voices without transparency or accountability.

- **A.I.-Driven Content Moderation and Censorship**: Social media platforms like Facebook, Twitter, and YouTube relied heavily on A.I.-driven algorithms for content moderation during the pandemic. While these tools were ostensibly used to counteract misinformation, they often ended up censoring legitimate debate and dissent. Terms of service and "community standards" were enforced selectively, leading to accusations that platforms were censoring information that ran counter to certain political or corporate agendas.

  - o For example, numerous users and even professional experts were de-platformed or had their content removed for questioning lockdown measures or expressing alternative viewpoints on public health responses. Critics argue that this selective enforcement amounts to social coercion, limiting individuals' ability to express views and access diverse perspectives in the public sphere (Bastos & Farkas, 2021).

- **Algorithmic Manipulation and Shadowbanning**: Algorithms on social media platforms are often designed in ways that prioritize certain content over others, impacting visibility and reach without users' knowledge or consent. **Shadowbanning** is a common tactic, where a user's content is made less visible or restricted without notifying them, effectively muting their voice without overt censorship. This practice has raised ethical concerns as it allows companies to suppress information or users based on subjective standards, creating a biased information ecosystem.

  o   For example, Twitter, Instagram, and Facebook have been criticized for allegedly shadowbanning users based on political affiliations or sensitive topics, creating an echo chamber where only certain narratives are prominent. Algorithmic manipulation also includes artificial promotion, where the visibility of certain users or companies is artificially boosted to present an inflated sense of popularity, social influence, or credibility. This selective amplification can distort public perception, making some individuals or viewpoints appear more legitimate or accepted than they actually are, undermining the authenticity of public discourse (Knight, 2021).

- **Ambiguous Community Guidelines as Tools of Control**: Many social media platforms use ambiguous community guidelines, allowing them to ban accounts, limit reach, or remove content without clear justification. These vague policies grant platforms significant discretion in enforcing rules, often with inconsistent applications. By keeping guidelines broad or undefined, companies can suppress or penalize certain users or companies selectively, depending on internal biases or external pressures.

  o   For instance, YouTube, Instagram, and Facebook have all been criticized for suspending or de-platforming users under the guise of "violating community standards" without providing specific reasons. This lack of transparency and selective enforcement can lead to what some describe as "exclusionary terrorism," where users and businesses are marginalized or silenced based on loosely defined terms. This ambiguity encourages self-censorship among users, who may avoid certain topics or viewpoints for fear of unintentional violations, thus stifling free expression (Sullivan, 2022).

5.3 Exclusionary Terrorism: Substandard Contact Methods to Frustrate Consumer Issues

Another method of what some consider "exclusionary terrorism" involves corporations intentionally creating poor customer service experiences to deter consumers from seeking support or redress for legitimate concerns. This tactic can lead to consumer frustration and ultimately dissuade individuals from addressing issues with the company, allowing the corporation to avoid accountability. Common tactics include:

- **Non-Functional or Obscure Contact Methods**: Some companies provide phone numbers that do not work or are difficult to find on their websites. This intentional obscurity can prevent consumers from reaching a real person, leaving them without a means to resolve their concerns.

- **Unresponsive or Delayed Email Responses**: Certain companies list email addresses as customer support channels but fail to respond in a timely manner, if at all. This tactic can exhaust consumers' patience, leaving issues unresolved without a clear avenue for follow-up.

- **Outsourced Customer Service with Limited Escalation Procedures**: Some corporations use foreign-based customer service agents who are given limited authority and are often unable to escalate calls to higher management when complex or legitimate issues arise. This lack of escalation procedures leaves consumers with limited options and can prevent them from obtaining meaningful assistance.

- **Automated Systems Without Human Alternatives**: Many companies increasingly rely on automated systems for customer service, such as chatbots or IVR (interactive voice response) systems, which often lack the sophistication to handle nuanced issues. This reliance on automation can be frustrating for consumers who wish to speak to a human representative, particularly for complex or sensitive concerns.

By implementing these exclusionary tactics, companies can frustrate consumers to the point where they abandon efforts to address legitimate grievances. This approach serves to minimize accountability and can allow companies to sidestep issues that may otherwise require reimbursement, corrective action, or public acknowledgment.

5.4 Long-Term Impacts and the Emboldening of Corporate Power Post-Lockdowns

Post-lockdowns, the influence of these corporations has only intensified. With their resources, data, and insights gained during the pandemic, these companies have become emboldened to assert even greater control over social and economic landscapes. The dependency on digital platforms established during the lockdowns has largely persisted, with tech giants now possessing unprecedented leverage over consumers, businesses, and even governments.

- **The Case of PayPal's Proposed Misinformation Fine**: In October 2022, PayPal faced intense backlash for a policy update that appeared to authorize the company to fine users up to $2,500 for spreading misinformation. The update was met with widespread public criticism, prompting PayPal to retract the policy and clarify that it was published in error. However, the incident raised significant concerns about corporate overreach and the potential for financial platforms to penalize users for subjective determinations of misinformation, thus exemplifying a form of social control through financial penalties (Yahoo Finance, 2022).

- **PayPal's Data Sharing Plans**: In 2023, PayPal announced plans to share customer data with third-party merchants, including Square, supposedly to enhance the shopping experience. Critics, however, questioned whether this was a strategy to cross-reference customer databases across platforms to exert more control over users' financial activities and personal preferences. This raised alarms about potential financial and social manipulation, as such data sharing could enable companies to surveil and possibly punish users based on behavioral patterns across different platforms, effectively enabling what some perceive as financial and social terrorism.

- **Consolidation of Market Power and Reduced Competition**: The pandemic accelerated trends toward digitalization, positioning a few corporations as gatekeepers of commerce, communication, and information. As small businesses were forced to adopt online sales, they often found themselves reliant on platforms operated by the same few tech giants. This reliance has allowed companies like Amazon, Google, and Apple to establish monopolistic or near-monopolistic control in their sectors, using their influence to dictate terms and reduce competition.

    o   Research indicates that between 2020 and 2022, large tech companies witnessed record profits and market consolidation, while smaller businesses either closed down or struggled to adapt. This concentration of economic power enables these corporations to shape industry standards, pricing, and consumer expectations, reinforcing a landscape where smaller competitors struggle to survive (Boushey, 2021).

5.5 Potential Countermeasures and Regulatory Recommendations

To address these risks of financial and social terrorism enabled by A.I., big data, and corporate dominance, governments and regulatory bodies need to implement policies that limit the unchecked power of large corporations over the economic and social fabric of society. Below are several proposed countermeasures:

- **Antitrust Measures and Market Competition Laws**: Stricter enforcement of antitrust laws is necessary to prevent monopolistic practices and protect competition. Breaking up large conglomerates or placing caps on the market share any one company can control would reduce the risk of economic manipulation by a few powerful entities. Europe, for example, has taken significant steps in this direction with the Digital Markets Act (DMA), which aims to curb the market power of "gatekeepers" in the tech sector.

- **Transparency in Data Collection and Usage**: Companies should be required to disclose their data collection practices, the specific uses of collected data, and provide consumers with the option to opt-out of certain types of data collection. Enhanced data transparency and user controls would empower individuals to better understand and control how their data is used, potentially reducing the scope for manipulation.

- **Algorithmic Accountability and Content Moderation Transparency**: Social media platforms should provide transparency regarding their content moderation policies, as well as insight into how A.I.-driven algorithms determine what content is prioritized or censored. Publicly available audits of algorithms could prevent misuse and provide greater accountability, helping ensure that social platforms don't restrict legitimate discourse or stifle dissent.

- **Support for Small Businesses and Digital Inclusivity**: Governments could provide incentives for small businesses to transition online without becoming dependent on large tech platforms. Digital inclusivity programs that support open-source technologies or independent e-commerce platforms would encourage a diverse digital ecosystem and reduce dependency on dominant corporations.

Conclusion of Chapter 5

The COVID-19 lockdowns and subsequent corporate actions highlighted the capacity of large corporations to wield A.I. and big data as tools of economic and social control, which some view as a form of financial and social terrorism. By capitalizing on their dominance during the pandemic, these companies entrenched themselves in a position of unprecedented influence over commerce, communication, and information. Without regulatory intervention, this influence is likely to continue expanding, creating risks for economic freedom, social equity, and democratic integrity.

Policymakers must address these concerns with targeted legislation aimed at curbing corporate power, enhancing transparency, and promoting competition. By implementing these measures, society can better guard against the risks of financial and social manipulation, ensuring that technology serves the public good rather than corporate interests alone.

References

1. Bastos, M., & Farkas, J. (2021). *COVID-19 and Conspiracy: Effects of Digital Information on Beliefs in COVID-19 Misinformation.* Social Media + Society.

2. Boushey, H. (2021). *Unbound: How Inequality Constricts Our Economy and What We Can Do About It.* Harvard University Press.

3. Knight, W. (2021). *The Invisible Influence of Algorithms and Shadowbanning.* MIT Technology Review. Retrieved from https://www.technologyreview.com/

4. Levy, S. (2021). *Facebook: The Inside Story.* Penguin Books.

5. Sullivan, M. (2022). *How Social Media Platforms Use Ambiguous Community Guidelines to Silence Dissent.* The Verge. Retrieved from https://www.theverge.com/

6. Soper, S. (2020). *Amazon's Monopolistic Practices During COVID-19 Pandemic.* Bloomberg.

7. Vox. (2022). *The Twitter Files: A Controversial Look Inside Twitter's Content Moderation.* https://www.vox.com/policy-and-politics/2022/12/15/23505370/twitter-files-elon-musk-taibbi-weiss-covid

8. Yahoo Finance. (2022). *PayPal Faces Backlash Over Proposed Misinformation Fines.* https://finance.yahoo.com/news/paypal-says-never-intended-fine-150420970.html

9. Bloomberg. (2020). *Amazon's Pandemic Power Surge and Small Business Challenges.* https://www.bloomberg.com/

10. Partnership on AI. (2021). *About Us.* https://www.partnershiponai.org/about/

CHAPTER 6: PRIVACY, AUTONOMY, AND CIVIL RIGHTS

6.1 Privacy Violations by A.I.-Enabled Surveillance Systems

A.I.-driven surveillance systems have become increasingly pervasive, posing significant threats to individual privacy. Governments, corporations, and other entities use these technologies for purposes ranging from public safety to consumer profiling, often without individuals' informed consent. The combination of facial recognition, behavior analysis, and location tracking allows organizations to monitor and record people's actions on an unprecedented scale, eroding the concept of personal privacy.

- **Facial Recognition and Identity Tracking**: Facial recognition technology, widely adopted in public and private spaces, can identify individuals in real-time, often without their knowledge. This technology is commonly used by law enforcement agencies, retail businesses, and airports for security purposes. However, facial recognition systems raise concerns about mass surveillance, potential misuse, and the tracking of individuals without due cause.

    o   In some countries, facial recognition systems have been deployed to monitor political protests or public gatherings, raising concerns about freedom of assembly and the potential chilling effect on civil rights (Smith & Miller, 2020). Without regulation, facial recognition could be used to suppress dissent and monitor political opponents, infringing upon democratic principles.

- **Consumer Profiling and Behavioral Data Collection**: Corporations often use A.I. and big data to analyze consumer behavior, creating highly detailed profiles for targeted advertising. These profiles include purchasing habits, browsing history, and personal interests, allowing companies to deliver personalized marketing. However, this data collection is frequently done without explicit user consent, raising questions about the boundaries of privacy.

o   The European Union's General Data Protection Regulation (GDPR) has set a precedent for data privacy, requiring companies to obtain explicit consent before collecting and processing personal data. GDPR represents a model for privacy protection, but similar protections are not yet standard worldwide.

6.2 Consent, Autonomy, and Individual Freedom

The rise of A.I. and automated decision-making systems has implications for personal autonomy and individual freedom. When algorithms influence decisions in areas such as finance, employment, and healthcare, individuals may feel a loss of control over aspects of their lives, especially when these decisions lack transparency or offer no clear avenue for appeal.

- **Algorithmic Decision-Making in High-Stakes Areas**: Automated decision-making is becoming commonplace in fields like hiring, credit scoring, and medical diagnostics. However, these decisions can feel opaque, with individuals often unaware of the factors influencing the outcomes. When individuals are denied a job, a loan, or a medical treatment based on an algorithm, they may feel that their autonomy has been compromised by faceless systems.

  o   Regulatory approaches such as the European Union's "Right to Explanation" in GDPR grant individuals the right to understand how decisions affecting them were made by automated systems. This principle could serve as a model for empowering individuals to challenge or appeal algorithmic decisions that affect their lives.

- **The "Nudging" Effect of A.I.-Driven Recommendations**: Many A.I. systems, particularly in digital platforms and e-commerce, are designed to nudge users towards specific actions or purchases based on behavioral data. These nudges are typically driven by algorithms that anticipate user preferences, but they can limit individual freedom by reinforcing existing behaviors and discouraging independent decision-making.

  o   While nudging can improve user experience and efficiency, it also raises ethical questions about manipulation and autonomy. Users may feel steered towards certain choices, often without realizing that A.I. is influencing their behavior. Greater transparency around recommendation algorithms could help restore agency, allowing individuals to make more informed decisions.

6.3 Balancing Security with Privacy Rights

The security benefits of A.I.-driven surveillance and data analytics are often cited to justify intrusions into individual privacy. While these technologies can enhance public safety, they must be balanced against the right to privacy to avoid creating a surveillance state that infringes upon civil rights and freedom.

- **The Security-Privacy Trade-Off**: The use of A.I. in law enforcement, border security, and public health surveillance has created a delicate trade-off between security and privacy. For example, during the COVID-19 pandemic, some countries implemented contact tracing apps to monitor and control the virus's spread. However, these systems collected extensive location and health data, leading to concerns about long-term data storage and potential misuse (Jones et al., 2021).

  o To achieve balance, governments must establish clear guidelines on data collection, retention, and usage for security purposes. Limiting data collection to the minimum required and implementing strong anonymization techniques can help protect privacy while still allowing for effective security measures.

- **The Role of Oversight and Accountability**: Independent oversight bodies can play a vital role in ensuring that security measures do not encroach on privacy rights. Establishing accountability mechanisms, such as regular audits, data access logs, and public reporting, can help maintain transparency and ensure that A.I.-enabled security measures are used responsibly. Judicial and legislative oversight is particularly important when sensitive personal data is involved.

6.4 Right to Choose Non-Engagement with A.I. Systems

As A.I.-driven services and devices become more integrated into daily life, individuals may increasingly encounter situations where interacting with A.I. is nearly unavoidable. This trend raises ethical concerns about the right to opt-out or refuse interaction with A.I. systems, especially in situations where personal data is collected or where decisions impact individual welfare.

- **Mandatory Use of A.I.-Driven Platforms**: Certain services and institutions now rely heavily on A.I.-driven systems, leaving users with limited options to avoid interaction. For example, job applications, bank transactions, and even some healthcare services are increasingly automated, making it difficult for individuals who prefer human interaction to opt-out of A.I.-based processes. Providing alternatives to A.I.-driven plat-forms is crucial to ensure that individuals retain the choice to interact with human representatives where feasible.

- **Privacy Concerns in Public Spaces**: As A.I. technologies like facial rec-ognition become standard in public and private spaces, the choice to avoid surveillance is nearly impossible. Individuals have little to no con-trol over being recorded or analyzed by A.I. systems in public, which raises concerns about the right to privacy in shared environments. Es-tablishing "A.I.-free" zones or giving individuals the option to avoid un-necessary surveillance could help mitigate these concerns.

6.5 Data Ownership and Algorithmic Transparency

Data ownership and transparency are central to discussions on A.I. governance and civil rights. Individuals generate vast amounts of data in their daily activi-ties, from browsing habits to geolocation information. However, they often lack control over how this data is collected, stored, and used by corporations and governments. Furthermore, the algorithms that process this data are frequently opaque, leaving users with little understanding of how A.I. systems influence their lives.

- **The Debate on Data Ownership**: Many argue that individuals should have greater ownership over the data they produce. Current models often place data ownership in the hands of companies collecting it, al-lowing them to profit from consumer data with minimal benefit to the users themselves. Data ownership rights could empower individuals to control who accesses their information and how it is used.

    o   In some jurisdictions, such as under the GDPR, individuals al-ready have limited data rights, such as the right to access, cor-rect, and delete their personal data. However, broader and more robust data ownership laws are needed to ensure individuals have a true say over their digital footprint.

- **Algorithmic Transparency and Explainability**: Transparency in A.I. al-gorithms is essential for building trust and accountability. Users should have the right to know how algorithms impacting their lives are de-veloped, the data they rely on, and the logic they employ. Algorithmic transparency can also help in detecting biases, ensuring fair treatment across different demographics.

o One approach is to require companies and government agencies to provide explanations for algorithmic decisions that significantly impact individuals, such as credit scoring, hiring, and legal sentencing. This "right to explanation" could be enforced through regulatory frameworks, enabling individuals to challenge or appeal decisions made by opaque A.I. systems.

Conclusion of Chapter 6

As A.I. systems permeate more aspects of society, safeguarding privacy, autonomy, and civil rights is paramount. While these technologies offer efficiency, convenience, and security, they must be implemented in ways that respect individual freedoms and prevent infringements on personal privacy. The right to privacy, freedom from algorithmic manipulation, and the ability to opt out of A.I. systems are fundamental aspects of digital citizenship in the modern world.

To protect these rights, regulatory measures focused on transparency, accountability, and data ownership are essential. Ensuring that A.I. and data-driven technologies are used responsibly will allow society to harness their benefits without compromising civil liberties. Policymakers must act decisively to uphold these principles, ensuring that technology enhances, rather than undermines, individual rights in the digital age.

References

1. Jones, M., Smith, P., & Miller, A. (2021). *The Impact of Contact Tracing on Privacy Rights During the COVID-19 Pandemic.* Journal of Public Health Policy.

2. Smith, A., & Miller, J. (2020). *Surveillance State: The Role of Facial Recognition in Policing and Privacy Concerns.* Journal of Law and Society.

3. Zuboff, S. (2019). *The Age of Surveillance Capitalism: The Fight for a Human Future at the New Frontier of Power.* PublicAffairs.

4. Binns, R. (2018). *Fairness in Machine Learning: Lessons from Political Philosophy.* Proceedings of the 2018 Conference on Fairness, Accountability, and Transparency.

5. Floridi, L. (2019). *The Ethics of Artificial Intelligence: Moral Perspectives on Human and Non-Human Agents.* Cambridge University Press.

CHAPTER 7: ENVIRONMENTAL AND RESOURCE MANAGEMENT

7.1 Energy Consumption of A.I. and Automation Systems

One of the most significant environmental concerns related to A.I. and automation technologies is the energy required to power advanced computing systems, especially in large-scale applications. Machine learning, particularly deep learning models, demands immense computational resources, which translates to high energy consumption. Data centers that support A.I. operations are energy-intensive and contribute to greenhouse gas emissions, making energy efficiency a priority for sustainable A.I. deployment.

- **Data Centers and Power Demand**: Data centers are the backbone of A.I. infrastructure, housing servers that process, store, and analyze vast amounts of data. These facilities require substantial energy for both computational processes and cooling systems to prevent overheating. Major tech companies are among the largest consumers of electricity globally, contributing significantly to carbon emissions (Masanet et al., 2020).

    o   To address this issue, companies are exploring renewable energy sources for data centers. Tech giants like Google and Microsoft have committed to operating their data centers on 100% renewable energy by set target dates. However, broader adoption of renewable energy across the industry remains challenging.

- **A.I. Model Training and Carbon Footprint**: Training large A.I. models, especially deep learning algorithms, consumes vast amounts of computational power. Studies indicate that training a single deep learning model can emit as much carbon as five cars over their entire lifespans (Strubell et al., 2019). This energy intensity underscores the need for more efficient A.I. architectures and algorithms that reduce environmental impact.

    o   Approaches to reducing energy consumption include optimizing model efficiency, using smaller datasets for pre-training, and adopting algorithms that require less computational power. By prioritizing energy-efficient A.I. models, researchers and companies can reduce their environmental footprint.

7.2 Resource Usage in A.I. Hardware and Manufacturing

The hardware required to run A.I. and automation systems includes servers, GPUs, TPUs, and other specialized components. Manufacturing these devices relies on finite resources, such as rare earth metals and minerals, which have significant environmental and social implications. Mining, extraction, and processing of these resources contribute to environmental degradation, and the demand for hardware components exacerbates this impact.

- **Rare Earth Metals and Mining Impact**: Components essential to A.I. hardware, such as lithium, cobalt, and nickel, are sourced from mining operations that often have adverse environmental effects. Mining processes can lead to deforestation, habitat loss, and water pollution. Moreover, mining is associated with poor labor conditions, particularly in developing countries (Manhart & Schleicher, 2015).

    o Reducing dependency on rare earth metals by investing in recycling programs, using alternative materials, and developing hardware that lasts longer are crucial steps in minimizing environmental impact. Companies must also adhere to ethical sourcing practices to ensure responsible procurement of materials.

- **Sustainable Hardware Design**: Designing hardware with environmental impact in mind can help reduce the resource intensity of A.I. and automation systems. This includes developing energy-efficient chips, optimizing hardware for durability and recyclability, and creating modular devices that are easier to repair or upgrade rather than replace. Sustainable hardware design not only reduces waste but also conserves the resources required for new production.

7.3 Disposal and Recycling of A.I. Systems

The disposal of outdated or non-functional A.I. hardware presents significant environmental challenges. E-waste, or electronic waste, is a growing problem as electronics reach the end of their lifecycle and are discarded. Without proper disposal and recycling practices, toxic materials in electronic components can leach into soil and water, posing long-term environmental risks.

- **E-Waste and Toxic Materials**: A.I. systems contain various materials that, if not properly disposed of, can harm the environment. Metals like lead, cadmium, and mercury, along with other hazardous chemicals, are common in electronic devices. When improperly disposed of, these toxins can contaminate groundwater and pose health risks to humans and wildlife.

  o Regulatory frameworks, such as the EU's Waste Electrical and Electronic Equipment (WEEE) Directive, mandate the safe disposal and recycling of electronic waste. Similar regulations should be adopted globally to ensure the responsible disposal of A.I. hardware, along with incentives for manufacturers to create more environmentally friendly products.

- **Circular Economy and Resource Recovery**: A circular economy approach aims to keep resources in use for as long as possible through reuse, repair, refurbishment, and recycling. Establishing efficient recycling programs for A.I. hardware can allow valuable materials to be recovered and repurposed, reducing the need for new resource extraction. Companies should invest in programs that facilitate product take-back, repair services, and recycling to support a more sustainable lifecycle for their devices.

7.4 Minimizing Resource Strain and Ecological Footprint

As A.I. and automation become more integrated into society, it's essential to consider their ecological footprint and the resources required to sustain them. Beyond energy and hardware, A.I. systems require extensive data storage and processing capabilities, which can strain physical and environmental resources.

- **Data Efficiency and Processing Optimization**: Developing algorithms that are data-efficient can reduce the amount of processing power and storage space required for A.I. operations. Techniques such as transfer learning, model compression, and federated learning reduce the resource demands of A.I. training and inference, making these processes more sustainable.

  o For example, federated learning enables A.I. models to be trained locally on devices instead of sending data to a central server, reducing bandwidth and storage requirements. Data-efficient algorithms are crucial for minimizing the ecological footprint of A.I. technologies, particularly as data generation continues to grow exponentially.

- **Cloud Computing and Environmental Efficiency**: Cloud computing providers, such as Amazon Web Services, Google Cloud, and Microsoft Azure, have significant environmental impacts due to the energy demands of their data centers. However, these companies are also in a position to lead in sustainability by adopting green energy sources, optimizing cooling technologies, and reducing water usage.

  o By centralizing computing resources, cloud providers can manage resource usage more efficiently than individual organizations. Encouraging the use of environmentally optimized cloud services over on-premises data centers can reduce the overall resource strain of A.I. technologies.

7.5 Environmental Protections and Regulations for A.I. Technologies

Regulations that address the environmental impact of A.I. systems are essential for sustainable development. Governments, in collaboration with industry stakeholders, can implement policies that promote environmental accountability, encourage sustainable practices, and provide incentives for adopting green technologies.

- **Regulatory Standards for Energy and Resource Efficiency**: Governments should establish standards that limit the energy consumption of A.I. data centers and require companies to use renewable energy sources. Incentives for energy-efficient data center design, such as tax breaks or grants, can encourage companies to prioritize sustainability in their operations.

- **Product Stewardship and Manufacturer Responsibility**: Manufacturers should be held accountable for the environmental impact of their products through product stewardship regulations. Extended producer responsibility (EPR) laws, which hold companies accountable for the lifecycle of their products, can incentivize sustainable design and ensure proper disposal and recycling practices.

  o In addition to EPR laws, governments can encourage innovation by providing funding for research in sustainable hardware design, efficient A.I. algorithms, and eco-friendly manufacturing methods.

- **Global Collaboration on A.I. Sustainability Goals**: Environmental issues related to A.I. are global in scope and require international collaboration. Establishing global sustainability goals, much like the United Nations' Sustainable Development Goals (SDGs), for A.I. technologies can encourage countries to work together on reducing the environmental impact of A.I. and promoting resource efficiency. Cross-border initiatives could include sharing best practices, developing joint recycling programs, and setting shared standards for data center energy efficiency.

Conclusion of Chapter 7

The environmental impact of A.I. and automation technologies presents a significant challenge as society seeks to balance innovation with sustainability. By addressing energy consumption, resource usage, waste management, and adopting responsible regulatory frameworks, the industry can mitigate its ecological footprint. Sustainable A.I. practices not only benefit the environment but also promote long-term viability and ethical responsibility within the field.

Policymakers, corporations, and research institutions must collaborate to advance sustainable practices in A.I. and automation. Through concerted efforts, it is possible to reduce the environmental burden of these technologies and contribute to a greener, more sustainable future.

References

1. Masanet, E., Shehabi, A., Lei, N., Smith, S., & Koomey, J. G. (2020). *Recalibrating Global Data Center Energy-Use Estimates*. Science, 367(6481), 984–986.

2. Strubell, E., Ganesh, A., & McCallum, A. (2019). *Energy and Policy Considerations for Deep Learning in NLP*. Proceedings of the 57th Annual Meeting of the Association for Computational Linguistics.

3. Manhart, A., & Schleicher, T. (2015). *The Social and Environmental Impact of Mining in Developing Countries*. Journal of Resources Policy.

4. Zuboff, S. (2019). *The Age of Surveillance Capitalism: The Fight for a Human Future at the New Frontier of Power*. PublicAffairs.

5. Floridi, L. (2019). *The Ethics of Artificial Intelligence: Moral Perspectives on Human and Non-Human Agents*. Cambridge University Press.

CHAPTER 8: REGULATORY AND GOVERNANCE FRAMEWORK

8.1 Proposed Global A.I. Regulatory Body

A.I. technology transcends borders, making it crucial to establish an international regulatory body to ensure consistent standards and ethical guidelines worldwide. This regulatory body would work in collaboration with national governments to address shared challenges, such as privacy, security, accountability, and ethical issues. Establishing a global framework would help prevent regulatory gaps that could allow companies or governments to exploit less regulated regions.

- **International Standards and Cooperation**: The proposed regulatory body would work to establish international standards for A.I. systems, ensuring that ethical considerations and safety protocols are consistent across countries. It would coordinate efforts among nations, creating a cooperative environment for managing global challenges such as cyber threats, data privacy, and ethical accountability.

  - Organizations like the United Nations (U.N.), the Organisation for Economic Co-operation and Development (OECD), and the European Union (EU) have taken initial steps in creating frameworks for digital governance. Leveraging these existing institutions or establishing a specialized international A.I. body would facilitate cooperation and standardization.

- **Ethics and Human Rights Oversight**: A central mission of the global regulatory body would be to monitor A.I. applications for potential human rights violations. For example, surveillance technologies used by governments or corporations could infringe on civil liberties if left unchecked. The regulatory body would establish ethical guidelines and accountability mechanisms to ensure that A.I. technologies respect human rights across all jurisdictions.

8.2 National vs. International A.I. Laws and Coordination

While a global framework is essential, national governments also need to develop and enforce their own A.I. regulations tailored to their specific legal and cultural contexts. Balancing national autonomy with international standards is necessary to foster compliance and ensure that regulations address both local and global concerns.

- **Local Regulations for Societal Needs**: Each nation faces unique challenges regarding privacy, security, and labor implications due to A.I. automation. For instance, countries with stricter privacy norms, such as those in the European Union, may require more stringent data protection measures than others. National laws can address specific societal concerns while still adhering to international principles.

  o The EU's General Data Protection Regulation (GDPR) serves as a model of national-level legislation with global implications, setting standards for data privacy that impact multinational corporations. Similarly, individual nations can enact laws that align with international standards while addressing their citizens' unique needs.

- **Coordination Mechanisms for Cross-Border Data Flow**: Since data often crosses international borders, effective coordination mechanisms are essential for managing data-related challenges, such as privacy and cybersecurity, across jurisdictions. Countries can establish treaties or data-sharing agreements to enable lawful data exchange while maintaining protection and accountability.

  o For example, the EU-U.S. Privacy Shield agreement (now under review and replaced by the Trans-Atlantic Data Privacy Framework) aimed to enable data flows between the EU and U.S. while ensuring that U.S. companies adhered to EU privacy standards. Similar agreements could foster data cooperation between countries with differing regulations.

8.3 Mechanisms for Compliance, Monitoring, and Enforcement

To ensure adherence to A.I. regulations, robust mechanisms for compliance, monitoring, and enforcement must be established. These mechanisms should involve regular audits, accountability protocols, and avenues for public reporting, empowering regulatory bodies to take corrective action if violations occur.

- **Regular Audits and Reporting Requirements**: Companies and institutions using A.I. in high-stakes or public-facing roles should undergo regular audits to ensure compliance with ethical and regulatory standards.

These audits would verify adherence to principles like fairness, transparency, and security. Requiring companies to submit periodic reports on A.I. activities can improve accountability.

> o The auditing process could involve a combination of self-assessments and independent third-party audits. This approach would ensure that companies are monitoring their own practices while remaining subject to external scrutiny.

- **Public Reporting and Whistleblower Protections**: Establishing a public reporting mechanism would allow individuals to report unethical or illegal uses of A.I. technology without fear of retaliation. Whistleblower protections would encourage employees and other stakeholders to come forward if they witness non-compliance, thus helping regulatory bodies detect issues that may otherwise go unnoticed.

- **Enforcement and Penalties**: Governments and regulatory bodies need the authority to enforce A.I. laws and impose penalties for non-compliance. Penalties could range from fines to restrictions on business operations, depending on the severity of the violation. For multinational companies, penalties could be coordinated across jurisdictions to prevent companies from relocating to avoid sanctions.

8.4 Licensing and Oversight for Development of Synthetic Humans and Advanced A.I. Systems

For advanced A.I. applications, such as synthetic humans and autonomous decision-making systems, additional licensing and oversight requirements may be necessary. These systems often operate in high-stakes environments, where their decisions have direct implications for human welfare, requiring a more rigorous regulatory approach.

- **Licensing for High-Risk A.I. Applications**: Companies developing synthetic humans, autonomous vehicles, or A.I.-powered healthcare diagnostics should obtain special licenses demonstrating compliance with safety and ethical standards. Licensing would establish a regulatory checkpoint to ensure these systems meet predefined criteria before they are deployed publicly.

> o For instance, autonomous vehicles undergo extensive testing and regulatory review to ensure they meet safety standards. A similar approach could be taken for A.I. systems in critical areas

like healthcare, law enforcement, or finance, where unanticipated errors could lead to significant harm.

- **Real-Time Oversight and Monitoring in Sensitive Applications**: High-risk A.I. systems, such as those in law enforcement or military applications, require continuous oversight. Real-time monitoring allows regulatory bodies or designated oversight agencies to track system performance and intervene if the A.I. exhibits harmful behavior. This is particularly important for A.I. with lethal or surveillance capabilities, where mistakes can lead to catastrophic outcomes.

- **Ethical Review Boards for Development of Synthetic Humans**: The development of synthetic humans or A.I. systems with complex decision-making capabilities should be subject to ethical review by independent boards. These boards would evaluate the ethical implications, societal impact, and potential risks associated with deploying such systems, ensuring they align with human rights and societal values.

8.5 Establishing A.I. Safety Standards and Emergency Shutdown Protocols

To prevent unintended consequences and manage A.I.-related crises, establishing safety standards and emergency shutdown protocols is essential. These measures would provide a safeguard against A.I. systems acting unpredictably or failing catastrophically, especially in applications where human safety is at risk.

- **A.I. Safety Standards for Reliability and Robustness**: A.I. safety standards should focus on ensuring system reliability, robustness, and fault tolerance. Systems deployed in high-stakes environments must undergo rigorous testing to confirm they can operate under a range of conditions without unexpected failure. These standards should also include guidelines for detecting and addressing bias or inaccuracies in A.I. models.

  o For example, the IEEE's Ethics in Action initiative and ISO standards for A.I. safety provide foundational principles for developing reliable and ethically sound A.I. These guidelines should be adopted as part of a broader regulatory framework to ensure safety across all sectors.

- **Emergency Shutdown and Failsafe Mechanisms**: All high-risk A.I. systems should include emergency shutdown protocols and failsafe mechanisms that allow operators to intervene if the system behaves unpredictably. This is especially important for autonomous systems in healthcare, transportation, and defense, where an unanticipated error could have life-or-death consequences.

o   For example, autonomous vehicles are typically designed with manual override systems that allow drivers to regain control in case of system malfunction. Similarly, A.I.-controlled surgical robots in healthcare must be equipped with emergency shutdown capabilities. Regulatory requirements for emergency protocols can help prevent catastrophic failures in critical A.I. applications.

Conclusion of Chapter 8

Building a comprehensive regulatory and governance framework for A.I. is crucial for balancing innovation with public safety, ethical integrity, and international cooperation. By establishing a global regulatory body, enacting national laws that align with these international standards, and creating mechanisms for compliance and accountability, governments can help ensure that A.I. technologies develop responsibly and ethically.

A multi-layered approach, encompassing licensing, oversight, and failsafe mechanisms, allows for responsible deployment of high-risk A.I. applications, especially those involving synthetic systems and autonomous decision-making. Through proactive governance and clear safety standards, society can harness the transformative potential of A.I. while safeguarding against misuse, unintended consequences, and ethical violations.

References

1.   Floridi, L. (2019). *The Ethics of Artificial Intelligence: Moral Perspectives on Human and Non-Human Agents*. Cambridge University Press.

2.   Russell, S., & Norvig, P. (2009). *Artificial Intelligence: A Modern Approach*. Prentice Hall.

3.   IEEE Global Initiative on Ethics of Autonomous and Intelligent Systems. (2020). *Ethically Aligned Design: A Vision for Prioritizing Human Well-being with Autonomous and Intelligent Systems*.

4.   Binns, R. (2018). *Fairness in Machine Learning: Lessons from Political Philosophy*. Proceedings of the 2018 Conference on Fairness, Accountability, and Transparency.

5.   European Commission. (2021). *Proposal for a Regulation Laying Down Harmonised Rules on Artificial Intelligence (Artificial Intelligence Act)*. Brussels: European Union.

CHAPTER 9: EDUCATION, PUBLIC AWARENESS, AND INCLUSIVITY

9.1 Promoting Public Awareness of A.I. Capabilities and Limitations

The rapid development of A.I. has created a gap between public understanding and the reality of A.I.'s capabilities and limitations. While A.I. can transform various sectors, there are misconceptions and fears surrounding its impact. Educating the public on what A.I. can—and cannot—do is essential to foster realistic expectations, reduce unnecessary fears, and increase societal acceptance.

- **Demystifying A.I.:** Many people view A.I. as a mysterious, almost magical technology. This lack of understanding fuels misconceptions and unwarranted fears, especially around job displacement, privacy, and surveillance. Public education campaigns and accessible resources explaining how A.I. works, its applications, and its limitations can help demystify the technology.

  - o   Workshops, seminars, and interactive online platforms hosted by universities, tech companies, and non-profits could offer practical information. These efforts should include clear explanations of terms like "machine learning," "deep learning," and "automation," showing where A.I. adds value and where human skills remain irreplaceable.

- **Managing Expectations About A.I. and Automation**: While A.I. can automate tasks and provide insights, it is far from achieving general intelligence or the full autonomy that some media portrayals suggest. Educating the public about these limitations can help set realistic expectations, reducing the fear of an A.I.-dominated future and highlighting A.I. as a complementary, not replacement, tool for human capabilities.

9.2 Education Initiatives on Ethics, Rights, and Interaction with A.I. Systems

Understanding A.I. extends beyond its technical aspects; it includes ethical implications, individual rights, and best practices for interacting with A.I. systems. As A.I. becomes a part of daily life, individuals should know their rights in relation to A.I., the ethical concerns surrounding its use, and how to navigate A.I.-enabled services confidently.

- **Teaching Digital Citizenship and Data Rights**: A key area of A.I. education should include digital citizenship and an understanding of personal data rights. Many individuals are unaware of the extent of data collected by A.I. systems, how it is used, and what rights they have to control this data. Teaching data literacy—particularly to young people—can empower individuals to make informed choices about their interactions with A.I. systems.

  o Schools, universities, and community organizations can integrate data literacy into their curricula. Resources explaining concepts like data privacy, consent, and algorithmic transparency would equip people with the knowledge to safeguard their digital rights.

- **Ethics and Responsible Use of A.I.**: Public education should include discussions on the ethical considerations of A.I., such as fairness, transparency, and accountability. Programs on A.I. ethics can highlight both the benefits and potential risks of A.I., encouraging informed and responsible usage. These discussions should address issues like bias in algorithms, privacy, and the human-A.I. relationship.

  o For example, a course or community program could use real-life case studies to illustrate ethical issues, exploring how A.I. decisions impact areas like hiring, healthcare, and law enforcement. By understanding these implications, people will be better prepared to critically assess and advocate for ethical practices in A.I.

9.3 Inclusive Policies to Ensure Fair Treatment of All Demographics

Ensuring that A.I. benefits society as a whole requires policies that promote inclusivity. A.I. systems must be developed and deployed in ways that consider all demographics, preventing biases and ensuring equitable access. Inclusive policies are necessary to avoid marginalizing certain groups and to ensure that everyone has the opportunity to benefit from A.I.-enabled progress.

- **Addressing Algorithmic Bias**: One of the most pressing concerns in A.I. ethics is algorithmic bias, which can result in discrimination against certain demographic groups. For instance, biased algorithms in hiring or lending decisions can disproportionately impact people of color, women, or economically disadvantaged individuals. Creating A.I. systems that consider diverse perspectives and data inputs can help reduce these biases.

o   Inclusive A.I. design requires a diverse team of developers, data scientists, and stakeholders. Governments and organizations can incentivize diversity in A.I. development, ensuring that multiple perspectives are considered in A.I. training and implementation.

- **Ensuring Equal Access to A.I. Benefits**: Socioeconomic disparities mean that certain populations may have less access to A.I.-enabled resources, from healthcare diagnostics to educational tools. Bridging this gap requires policies that ensure equitable access to A.I. benefits across income levels and geographic regions.

o   Public sector investment in A.I.-enabled resources for underserved communities, like rural areas or low-income neighborhoods, can promote inclusivity. For example, using A.I. in public health could help detect and address medical needs in communities with limited healthcare access.

9.4 Addressing Misinformation and Fearmongering

Misinformation and fearmongering around A.I. often fuel unrealistic fears and resistance, which can slow down beneficial advancements. By addressing common misconceptions and promoting fact-based discussions, society can build a more balanced view of A.I. technologies and their impact.

- **Combating A.I.-Related Misinformation**: False information about A.I. technologies—such as the belief that A.I. will imminently replace all jobs or take control over humans—can lead to unnecessary fears. Governments, academic institutions, and media organizations can collaborate to promote factual information about A.I., focusing on the technology's current state and realistic future.

o   Initiatives like fact-checking resources, public service announcements, and partnerships with credible institutions could serve to counteract A.I.-related misinformation. Additionally, platforms could label and flag misleading content on social media, ensuring the public has access to accurate A.I. information.

- **Encouraging Open Dialogues and Public Forums**: Public forums, workshops, and discussion panels about A.I. can help people voice their concerns, ask questions, and gain a better understanding of how A.I. will impact society. Transparent dialogues on A.I. allow experts to address public anxieties while educating on the technology's benefits.

    o   Regularly hosted panels or community workshops, with participation from A.I. experts, ethicists, and community leaders, could create spaces for nuanced discussion. These events can address current developments, ethical concerns, and the potential social impact of A.I., fostering a balanced and informed public perspective.

9.5 Encouraging Global Dialogue on the Role of A.I. in Society

Given the global nature of A.I. development and its wide-reaching impact, an international dialogue on the societal role of A.I. is essential. Different cultures and societies may have varying attitudes towards A.I., but establishing shared principles can promote responsible and ethical development.

- **International Collaborations on A.I. Ethics and Governance**: Countries around the world face similar ethical and social challenges posed by A.I., from privacy concerns to job displacement. Collaborative efforts, such as international conferences and partnerships, allow for sharing insights, best practices, and regulatory approaches that promote ethical A.I. deployment.

    o   Initiatives like the Partnership on AI, a consortium involving stakeholders from multiple countries, offer a model for cross-border collaboration. These platforms enable countries to address common concerns while considering cultural and regulatory differences.

- **Promoting Cultural Sensitivity in A.I. Development**: A.I. systems deployed internationally should consider local cultural norms, legal requirements, and ethical standards. Promoting culturally aware A.I. deployment ensures that the technology aligns with the values of different societies, helping prevent cultural or ethical clashes.

    o   For instance, A.I. systems used in public surveillance may be accepted in one country but raise concerns in another due to privacy expectations. International A.I. standards should account for such cultural variations, promoting respectful and responsible use across borders.

Conclusion of Chapter 9

Education, public awareness, and inclusivity are essential to fostering a society that can responsibly navigate the challenges and opportunities posed by A.I. technologies. By investing in education initiatives, encouraging inclusivity in A.I. development, and promoting fact-based dialogues, society can demystify A.I., dispel fears, and ensure that all demographics have a stake in its benefits.

An informed public is better equipped to advocate for ethical and responsible A.I. deployment, balancing technological progress with human values. As A.I. becomes increasingly integrated into daily life, these efforts will be crucial in shaping a future where technology serves the public good, respects cultural diversity, and promotes social equity.

References

1.   Floridi, L. (2019). *The Ethics of Artificial Intelligence: Moral Perspectives on Human and Non-Human Agents.* Cambridge University Press.

2.   Eubanks, V. (2018). *Automating Inequality: How High-Tech Tools Profile, Police, and Punish the Poor.* St. Martin's Press.

3.   Noble, S. U. (2018). *Algorithms of Oppression: How Search Engines Reinforce Racism.* NYU Press.

4.   Partnership on AI. (2021). *About Us.* [Online] Available at: https://www.partnershiponai.org/about/

5.   Binns, R. (2018). *Fairness in Machine Learning: Lessons from Political Philosophy.* Proceedings of the 2018 Conference on Fairness, Accountability, and Transparency.

CHAPTER 10: FUTURE-PROOFING AND UNKNOWN THREATS

10.1 Preparing for Potential Unknown A.I. Risks

As A.I. technologies evolve, new risks may emerge that cannot be fully anticipated with today's knowledge. Preparing for these unknown threats involves creating flexible and adaptive policies that can evolve alongside technological advancements. Regulatory frameworks should be designed to handle a wide range of possible scenarios, from minor developments to significant breakthroughs that challenge current assumptions.

- **Flexible Regulatory Frameworks**: Static regulations can quickly become outdated as A.I. technology advances. By adopting a flexible regulatory framework, governments and institutions can adapt to new developments without requiring frequent legislative overhauls. This flexibility could involve setting up adaptive regulatory mechanisms that allow for periodic reviews and updates based on technological trends and ethical considerations.

  o A model framework could include sunset clauses in regulations, allowing them to expire or be revised within a set timeframe. Regular evaluation cycles, perhaps every five years, could ensure that policies stay relevant in the face of rapid A.I. evolution.

- **Establishing Multi-Disciplinary Task Forces**: To address unknown risks, multi-disciplinary task forces could be established, consisting of experts from fields such as computer science, ethics, law, psychology, and sociology. These task forces would monitor developments in A.I. and recommend proactive adjustments to policies, bridging the gap between technology and society's evolving needs.

  o An example of this is the U.K.'s Centre for Data Ethics and Innovation, which advises on ethical and innovative uses of data and A.I. Establishing similar bodies globally would help track and anticipate future A.I. challenges.

10.2 Scenarios for Rapid Technological Advancements

A.I. research and development could lead to rapid advancements that fundamentally alter its capabilities and applications. Future-proofing strategies must account for potential scenarios where A.I. exceeds current limitations, enabling new forms of intelligence, automation, or control.

- **Breakthroughs in General Artificial Intelligence (AGI)**: While today's A.I. is limited to narrow applications, future advancements could bring about general artificial intelligence (AGI) with cognitive capabilities comparable to human reasoning. AGI would be capable of learning and applying knowledge across various domains, raising questions about autonomy, control, and coexistence with humans.

    o   Policies for AGI should emphasize ethical considerations, including the need for strict safety and containment protocols. Establishing "alignment protocols," ensuring AGI goals align with human interests, would be crucial in maintaining AGI as a beneficial force.

- **Accelerated Development of Autonomous Systems**: Autonomous systems, such as self-driving vehicles or drones, could become increasingly sophisticated and widespread, operating independently with minimal human intervention. This expansion could present logistical and ethical challenges, particularly in high-stakes or military applications.

    o   Regulators should consider establishing zones for autonomous operations and develop protocols for managing A.I.-driven systems in public spaces. Policies around autonomous operations in military and critical infrastructure contexts are particularly important, as mistakes could lead to significant public safety risks.

10.3 Crisis Management and Contingency Planning

Contingency planning is essential for managing crises that may arise from A.I. failures or malicious use. The complex nature of A.I. systems means that even well-designed technologies could experience unpredictable issues that require immediate intervention. Establishing crisis management protocols ensures preparedness for these eventualities.

- **Incident Response Teams for A.I. Emergencies**: A.I.-specific incident response teams, comprising technologists, ethicists, cybersecurity experts, and crisis managers, should be established to handle emergencies involving A.I. systems. These teams would respond to incidents ranging from system malfunctions to cyberattacks on A.I.-enabled infrastructure.

  o Incident response teams would perform rapid risk assessment and take corrective actions, such as shutting down compromised systems or enacting containment measures. These teams would also be responsible for investigating the root cause of the incident and recommending safeguards to prevent future occurrences.

- **Contingency Planning for A.I. Malfunctions**: In cases where critical A.I. systems, such as autonomous vehicles or healthcare robots, malfunction, rapid response protocols are crucial. Emergency shutdown mechanisms and manual override systems should be in place to allow human intervention if an A.I. system exhibits dangerous or unintended behavior.

  o Simulation exercises and drills involving A.I. systems can prepare operators for real-life contingencies. These exercises would test response protocols, ensuring that operators know how to deactivate or redirect A.I. systems in case of a malfunction or emergency.

10.4 Future-Proofing Policies to Adapt to Emerging Technologies

As A.I. technology continues to evolve, future-proofing policies should ensure adaptability for various advancements, including unanticipated innovations. Policymakers should focus on broad principles that can apply to a wide array of emerging technologies, allowing them to keep pace with changes without requiring frequent regulatory revisions.

- **Principle-Based Regulation**: Principle-based regulation establishes overarching principles that apply to any new development rather than focusing on specific technologies. This approach enables policymakers to enforce general standards, such as transparency, accountability, and fairness, which remain relevant regardless of how A.I. evolves.

    o   For instance, principles around privacy protection would remain applicable even as new forms of data analysis and collection methods emerge. By anchoring regulations in ethical principles, future policies can address innovations in a consistent, adaptable manner.

- **Regularly Updated Ethical Guidelines**: Ethical guidelines for A.I. should be periodically reviewed and updated to reflect societal values and technological capabilities. As A.I. systems begin to impact new areas of life, ethics boards should assess and address emerging concerns. Ethical reviews could cover issues like transparency, bias, and the societal impact of automation.

    o   Bodies such as the IEEE or AI Now Institute periodically publish updated guidelines for ethical A.I. development, providing a model for continuous ethical review. Integrating these guidelines into policy frameworks would help ensure that A.I. remains aligned with societal values over time.

10.5 Ethical Considerations for Unforeseen Capabilities of A.I. Systems

As A.I. systems gain new capabilities, ethical considerations will evolve to address questions around autonomy, rights, and societal impact. The rapid pace of A.I. innovation means that ethical frameworks must be flexible, balancing the benefits of progress with concerns over human welfare, justice, and fairness.

- **Addressing A.I. Autonomy and Control**: If future A.I. systems gain higher degrees of autonomy, managing control over their behavior and decision-making will become critical. Policies must ensure that autonomous systems remain accountable to human oversight, with clear protocols for intervening if systems act against human interests.

    o   Maintaining "human-in-the-loop" requirements, where human operators can supervise and override A.I. decisions, can help maintain control over autonomous systems. This approach ensures that humans retain ultimate authority over A.I. systems in high-stakes contexts.

- **Rights and Welfare of Synthetic Entities**: In the unlikely event that A.I. develops advanced cognitive capabilities or consciousness, discussions about the rights and welfare of synthetic entities may arise. If synthetic entities gain characteristics resembling sentience, ethical considerations around their treatment and responsibilities could become relevant.

o   While this remains speculative, establishing ethical principles around synthetic entities could prevent future dilemmas. For example, policies might address the use of highly advanced synthetic entities in labor or companionship roles, ensuring their welfare if they exhibit characteristics associated with self-awareness.

Conclusion of Chapter 10

Future-proofing policies for A.I. involve creating flexible frameworks that can adapt to rapid technological advancements and unforeseen threats. By establishing adaptive regulations, crisis management protocols, and ethical principles, society can better prepare for both the anticipated and unknown implications of A.I. development.

This approach to A.I. governance emphasizes resilience and adaptability, ensuring that society is prepared to harness the benefits of A.I. technologies while minimizing potential risks. In a world of rapid innovation, these future-proofing measures will be essential for creating a sustainable and ethical relationship between humans and artificial intelligence.

References

1.   Floridi, L. (2019). *The Ethics of Artificial Intelligence: Moral Perspectives on Human and Non-Human Agents*. Cambridge University Press.

2.   IEEE Global Initiative on Ethics of Autonomous and Intelligent Systems. (2020). *Ethically Aligned Design: A Vision for Prioritizing Human Well-being with Autonomous and Intelligent Systems*.

3.   Russell, S., & Norvig, P. (2009). *Artificial Intelligence: A Modern Approach*. Prentice Hall.

4.   Binns, R. (2018). *Fairness in Machine Learning: Lessons from Political Philosophy*. Proceedings of the 2018 Conference on Fairness, Accountability, and Transparency.

5.   AI Now Institute. (2020). *Ethics in Artificial Intelligence: Developing Policies for Rapid Technological Change*.

CHAPTER 11:
ETHICAL HUMAN LEADERSHIP
IN A.I. GOVERNANCE

11.1 Importance of Human Leadership in A.I. Oversight

As A.I. systems continue to impact all aspects of human life, the role of ethical human leadership becomes increasingly vital. While technological expertise is essential for overseeing A.I. development, leaders with a proven track record of advocating for humanity—through social causes, community service, and protection of rights—bring an irreplaceable moral perspective. These individuals can ensure that A.I. policies prioritize human welfare and that technology serves as a tool for societal advancement, not exploitation.

Leaders who have consistently championed human rights, social equity, and environmental sustainability bring a depth of understanding about human needs, struggles, and values that A.I. cannot replicate. Their involvement in A.I. governance can help maintain a human-first approach, emphasizing empathy, inclusivity, and fairness in A.I.-driven decision-making.

11.2 Identifying and Selecting Ethical Leaders
with a Humanity-First Approach

Not all leaders in technology are automatically suited for roles in A.I. governance. Selecting ethical leaders requires identifying individuals who have demonstrated their commitment to humanity through concrete actions. Qualities to consider include:

- **Social Advocacy and Humanitarian Work**: Leaders who have been actively involved in addressing social issues, such as homelessness, food security, and healthcare access, bring a perspective rooted in compassion and direct engagement with human needs.

- **Experience in Rights Protection and Fairness**: Those with a background in legal or social justice work can provide valuable insights into protecting individuals from exploitation, discrimination, and other harms. Their experience ensures that A.I. policies are designed with safeguards against abuse.

- **Commitment to Transparency and Accountability**: A history of advocating for transparency in governance—whether in corporate, non-profit, or governmental sectors—demonstrates a leader's dedication to accountability, an essential trait for ethical A.I. oversight.

11.3 Roles and Responsibilities of Human Advocates in A.I. Governance

Human advocates in A.I. governance should serve in roles that allow them to actively shape policies, monitor A.I. implementations, and hold corporations accountable for their actions. Responsibilities may include:

- **Establishing Ethical Standards and Review Processes**: These leaders should be involved in setting ethical standards for A.I. development, ensuring that core principles like fairness, transparency, and respect for individual rights are incorporated into A.I. frameworks.

- **Monitoring and Enforcing Compliance**: Ethical leaders should oversee compliance with ethical standards and intervene if corporations or institutions deviate from these guidelines. They should have authority to conduct audits and enforce penalties if A.I. implementations fail to meet ethical criteria.

- **Public Engagement and Education**: Human advocates should engage with the public to foster transparency around A.I. developments. By offering explanations, addressing concerns, and promoting public discourse, they can build trust between society and the entities developing A.I. technologies.

11.4 Case Studies of Ethical Leadership in Technology and Advocacy

To illustrate the importance of ethical human leadership, consider case studies of individuals or organizations who have successfully balanced technological advancement with human-centered advocacy. These case studies could include:

- **Tech CEOs who advocate for humanitarian causes**: Examples of leaders in the tech industry who have taken proactive stances on social issues, using their influence to promote positive societal changes.

- **Non-Profit Leaders in Technology for Good Initiatives**: Highlighting leaders from non-profits focused on using technology for social good, who bring a unique, value-driven approach to A.I. governance.

- **Legal Advocates in Technology Policy**: Individuals who have influenced A.I. legislation and regulatory frameworks, ensuring protections for marginalized communities and fostering responsible technology use.

11.5 Challenges and Limitations of Relying on Human Advocates

While ethical human advocates bring invaluable insights to A.I. governance, they face certain challenges, such as:

- **Conflicts of Interest and Biases**: Even well-intentioned leaders may have biases or face conflicts of interest, especially if they have affiliations with corporations or institutions that benefit from A.I. applications. Mechanisms to manage conflicts and ensure objectivity are essential.

- **Pressure from Profit-Driven Entities**: Advocates may encounter resistance from corporations or government bodies prioritizing economic gain over ethical concerns. Effective governance structures are necessary to empower ethical leaders to uphold human-first principles, even in the face of opposing interests.

- **Balancing Technological Progress with Ethical Concerns**: Human advocates must navigate the fine line between encouraging technological innovation and enforcing ethical limitations. Balancing these interests requires diplomacy, foresight, and a deep understanding of both technology and societal impact.

Conclusion of Chapter 11

Incorporating ethical human leaders into A.I. governance is essential to uphold values of fairness, transparency, and human welfare in a rapidly evolving technological landscape. By selecting individuals with a track record of advocating for humanity, A.I. oversight bodies can ensure that A.I. systems are developed and deployed responsibly, prioritizing societal good over profit or efficiency alone.

Such leaders act as a moral compass, guiding A.I. governance frameworks to foster a future where A.I. is a tool for enhancing human potential, not compromising it. With their oversight, society can work toward a future where A.I. and humanity coexist in harmony, promoting justice, equity, and shared prosperity.

CONCLUSION

Summarizing the Need for Ethical and Responsible A.I. Governance

As we have seen throughout this document, A.I. has the potential to profoundly influence nearly every aspect of human life. From improving healthcare outcomes and revolutionizing industries to enhancing public services, A.I. offers powerful tools to address some of society's most significant challenges. However, A.I. also introduces substantial risks, particularly when wielded by corporations and institutions that may prioritize profit, efficiency, or control over ethical considerations and public welfare.

This framework has outlined the many facets of responsible A.I. governance, highlighting the need for policies that go beyond mere regulatory compliance. True governance of A.I. requires a foundational commitment to ethics, human rights, and social responsibility. We must recognize that A.I., if left unchecked, has the potential to perpetuate discrimination, deepen social inequalities, and even erode the autonomy and rights of individuals. Thus, this document advocates for a governance model that prioritizes humanity at every stage of A.I. development and deployment, safeguarding our fundamental values in a rapidly evolving digital landscape.

Call to Action for Governments, Corporations, and Individuals

1. **Governments**: Policymakers must adopt proactive, adaptive regulations that keep pace with technological advancements while preserving democratic values and protecting citizens' rights. Governments must play a central role in setting ethical standards for A.I. and enforcing compliance, ensuring that A.I. technologies align with societal values. By fostering international collaboration, governments can establish consistent guidelines, closing regulatory gaps that could otherwise be exploited by multinational corporations.

2. **Corporations**: Companies at the forefront of A.I. development bear a profound responsibility to prioritize ethical design, transparency, and accountability. Corporations should voluntarily adopt practices that promote fairness, inclusivity, and the well-being of their users, rather than waiting for regulatory mandates. Self-regulation and a commit-

ment to humanity-first principles will not only protect the public but also enhance corporate reputation, building trust with consumers and society at large.

3. **Individuals**: Every individual has a role in shaping the trajectory of A.I. governance. By staying informed, advocating for transparency, and supporting ethical practices, citizens can influence both public policy and corporate behavior. Consumers and voters can demand accountability from both governments and companies, reinforcing a societal expectation that A.I. must serve the public good rather than narrow interests.

The Role of Ethical Human Leadership in Guiding A.I. Governance

One of the key additions to this framework is the emphasis on ethical human leadership within A.I. governance structures. Leaders with proven records of advocating for humanity, whether through social justice, community service, or the protection of individual rights, bring an essential moral perspective to A.I. oversight. Their commitment to empathy, transparency, and fairness helps ensure that A.I. policies are not merely reactive but rooted in a proactive dedication to human-centered values.

The presence of ethical leaders in A.I. governance can prevent misuse and abuse by ensuring that A.I. is guided by principles of justice, equality, and respect for human dignity. These individuals act as stewards of societal values, providing a necessary counterbalance to corporate and political interests that might prioritize efficiency or control over human welfare. By placing these ethical advocates at the heart of A.I. decision-making, we create a governance model that truly serves humanity.

Envisioning a Future with Ethical and Responsible A.I.

This framework envisions a future where A.I. and humanity coexist in a balanced, respectful partnership. In this vision, A.I. does not overshadow human agency but enhances our capabilities, addressing complex challenges and unlocking new opportunities for progress. By aligning A.I. with humanity's highest ethical standards, we can ensure that technology remains a tool for empowerment, not a force for domination.

To realize this vision, society must embrace a balanced approach that safeguards individual autonomy, promotes fairness, and values inclusivity. A.I. should be a driver of social equity, providing resources and opportunities to those who have historically been marginalized or disadvantaged. Rather than amplifying disparities, A.I. should help bridge gaps, creating a more just and compassionate world.

Confronting Risks: Financial and Social Terrorism, Privacy Concerns, and Exclusionary Tactics

Throughout this document, we have examined specific risks associated with A.I. misuse, such as financial and social terrorism, where corporations manipulate data and technology to control markets and public opinion. The potential for A.I.-enabled surveillance and exclusionary tactics, like limiting access to essential services or employing poor customer service to frustrate consumers, underscores the urgency of robust, ethical A.I. governance.

This framework calls for stringent measures to counteract these threats, including data transparency, algorithmic accountability, and policies that protect individual rights. Financial and social terrorism enabled by A.I. highlights the need for careful oversight, as corporate power can quickly become oppressive without checks and balances. Privacy, autonomy, and inclusivity must be safeguarded to protect individuals from exploitation and to ensure that A.I. remains a tool of liberation, not control.

Vision for a Harmonious Coexistence of Humans and Technology

In conclusion, this framework is more than just a set of guidelines; it is a vision for a future where humans and technology coexist harmoniously. It is a call for vigilance, responsibility, and compassion. A.I. is a powerful force, but without ethical grounding, it can become a tool of coercion and inequality. By committing to ethical human leadership, transparent practices, and a humanity-first approach, we can shape a world where A.I. serves as a beacon of progress, inclusivity, and equity.

As we move forward, let this framework inspire action among governments, corporations, and individuals. Let it remind us that we are not passive recipients of technology's impact but active shapers of its future. The choices we make today will define the legacy of A.I. for generations to come. May we choose wisely, embracing a future where A.I. is an extension of our shared values, driving positive change while honoring the diversity, dignity, and resilience that make us human.

Together, let us build an A.I.-enabled future that is just, inclusive, and prosperous—a future worthy of our highest aspirations.

USEFUL LINKS

Here's an extensive list of categorized links to provide you with resources relevant to each chapter, as well as additional information to deepen you understanding of A.I. regulation, ethics, environmental impact, and social implications. These links cover a range of organizations, governing bodies, publications, and educational resources that align with the chapters in the framework.

General Governing Bodies and International Organizations

1. **United Nations (U.N.)**
   - o United Nations Official Website: https://www.un.org/
   - o U.N. Global Pulse (A.I. and Big Data for Sustainable Development): https://www.unglobalpulse.org/

2. **Organisation for Economic Co-operation and Development (OECD)**
   - o OECD A.I. Policy Observatory: https://www.oecd.ai/
   - o OECD Digital Economy Policy: https://www.oecd.org/sti/digital-economy/

3. **European Union (EU)**
   - o European Commission – A.I. and Robotics: https://ec.europa.eu/digital-strategy/our-policies/artificial-intelligence_en
   - o EU General Data Protection Regulation (GDPR): https://gdpr.eu/

4. **World Economic Forum (WEF)**
   - o WEF Centre for the Fourth Industrial Revolution: https://www.weforum.org/centre-for-the-fourth-industrial-revolution
   - o WEF A.I. and Machine Learning: https://www.weforum.org/topics/artificial-intelligence-and-machine-learning/

Chapter 1: Ethical Considerations in A.I. Development

1. **Partnership on AI**
   - o Official Website: https://www.partnershiponai.org/

2. **IEEE Global Initiative on Ethics of Autonomous and Intelligent Systems**
   - o Ethically Aligned Design Document: https://ethicsinaction.ieee.org/

3. **AI Now Institute**
   o   Research on A.I. and Social Impacts: https://ainowinstitute.org/

4. **Center for Humane Technology**
   o   Promoting Ethical Tech Development: https://www.humanetech.com/

5. **United Nations – Artificial Intelligence for Good**
   o   A.I. for Good Global Summit: https://aiforgood.itu.int/

Chapter 2: Societal Impacts of A.I. and Automation

1. **World Economic Forum – Future of Jobs**
   o   Report on the Impact of A.I. on Jobs: https://www.weforum.org/reports/the-future-of-jobs-report-2020

2. **International Labour Organization (ILO)**
   o   Future of Work and A.I. Impact: https://www.ilo.org/global/topics/future-of-work/lang--en/index.htm

3. **European Union – Digital Skills and Jobs**
   o   EU Initiatives on Skills for the Digital Age: https://ec.europa.eu/digital-strategy/our-policies/digital-skills-jobs-coalition_en

4. **AI for Humanity (France)**
   o   National Strategy on A.I. Impact: https://www.aiforhumanity.fr/

Chapter 3: Technological Advancements and Capabilities

1. **MIT Technology Review**
   o   Latest in A.I. and Emerging Technologies: https://www.technologyreview.com/

2. **OpenAI**
   o   Research and Publications on A.I. Advancements: https://www.openai.com/

3. **Stanford Human-Centered Artificial Intelligence (HAI)**
   o   Research on A.I. Capabilities and Societal Impact: https://hai.stanford.edu/

4. **Google Research**
   o   Google's A.I. Research and Publications: https://research.google/

Chapter 4: Security Risks and Threat Management

1. **Center for Security and Emerging Technology (CSET)**

    o Security and A.I. Policy Research: https://cset.georgetown.edu/

2. **NATO Cooperative Cyber Defence Centre of Excellence (CCDCOE)**

    o Cybersecurity in A.I.: https://ccdcoe.org/

3. **RAND Corporation**

    o Research on A.I. and National Security: https://www.rand.org/topics/artificial-intelligence.html

4. **Cybersecurity & Infrastructure Security Agency (CISA)**

    o U.S. Government Cybersecurity Resources: https://www.cisa.gov/

Chapter 5: Financial and Social Terrorism

1. **European Central Bank (ECB)**

    o Digital Euro and Financial Technologies: https://www.ecb.europa.eu/home/html/index.en.html

2. **Financial Conduct Authority (FCA) – UK**

    o Financial Regulations and A.I. Impact: https://www.fca.org.uk/

3. **U.S. Federal Trade Commission (FTC)**

    o Consumer Protection and A.I.-Driven Financial Practices: https://www.ftc.gov/

4. **Global Financial Integrity (GFI)**

    o Research on Illicit Financial Flows and Technology: https://gfintegrity.org/

5. **MIT Technology Review – The Invisible Influence of Algorithms and Shadowbanning**

    o An article on algorithmic manipulation and the impact of shadowbanning: https://www.technologyreview.com/

6. **The Verge – Ambiguous Community Guidelines and Social Media Control**

    o Coverage on how platforms use vague guidelines to limit reach and control discourse: https://www.theverge.com/

7. **Vox – The Twitter Files Coverage**

   o Analysis of the "Twitter Files" Controversy and Its Implications: https://www.vox.com/policy-and-politics/ 2022/12/15/23505370/twitter-files-elon-musk-taibbi-weiss-covid

8. **Yahoo Finance – PayPal's Misinformation Fine Controversy**

   o Article on PayPal's Proposed $2,500 Fine for Misinformation: https://finance.yahoo.com/news/paypal-says-never-intend-ed-fine-150420970.html

9. **Bloomberg – Amazon's Market Practices During the Pandemic**

   o Overview of Amazon's Monopolistic Practices During COVID-19: https://www.bloomberg.com/

10. **Penguin Random House –**
    Facebook: The Inside Story by Steven Levy

    o Book on Facebook's Dominance and Data Practices: https://www.penguinrandomhouse.com/

11. **Partnership on AI**

    o Organization focused on responsible and ethical development of A.I. technology: https://www.partnershiponai.org/about/

Chapter 6: Privacy, Autonomy, and Civil Rights

1. **Electronic Frontier Foundation (EFF)**

   o Advocacy on Privacy and Digital Rights: https://www.eff.org/

2. **American Civil Liberties Union (ACLU)**

   o A.I., Privacy, and Civil Rights: https://www.aclu.org/issues/privacy-technology

3. **European Data Protection Board (EDPB)**

   o GDPR and Privacy Regulations in the EU: https://edpb.europa.eu/

4. **Future of Privacy Forum**

   o Data Privacy and A.I. Ethics Research: https://fpf.org/

Chapter 7: Environmental and Resource Management

1. **International Energy Agency (IEA)**

   o   Energy Usage in Technology and Data Centers:
       https://www.iea.org/

2. **Greenpeace – Click Clean Report**

   o   Data Centers and Renewable Energy:
       https://www.greenpeace.org/international/

3. **Natural Resources Defense Council (NRDC)**

   o   E-Waste and Sustainable Tech Practices:
       https://www.nrdc.org/

4. **Ellen MacArthur Foundation – Circular Economy**

   o   Circular Economy and Technology:
       https://www.ellenmacarthurfoundation.org/

Chapter 8: Regulatory and Governance Framework

1. **European Commission – Artificial Intelligence Act**

   o   Proposed Regulations on A.I. in the EU: https://ec.europa.eu/
       digital-strategy/artificial-intelligence-act_en

2. **Institute of Electrical and Electronics Engineers (IEEE)**

   o   Standards for A.I. and Governance: https://standards.ieee.org/
       initiatives/artificial-intelligence-systems/

3. **U.S. National Institute of Standards and Technology (NIST)**

   o   NIST A.I. Standards and Guidelines:
       https://www.nist.gov/artificial-intelligence

4. **Center for Data Ethics and Innovation (CDEI) – U.K.**

   o   Ethical A.I. Policy Development: https://www.gov.uk/
       government/organisations/centre-for-data-ethics-and-
       innovation

Chapter 9: Education, Public Awareness, and Inclusivity

1. **AI4ALL**

   o   Education Programs on A.I. for Youth and
       Underrepresented Groups: https://ai-4-all.org/

2. **Khan Academy – A.I. and Data Science**

   o   Free Courses on A.I. Basics and Data Science: https://www.khanacademy.org/

3. **Coursera – A.I. Ethics Courses**

   o   Online Courses on A.I. Ethics and Public Policy: https://www.coursera.org/

4. **IBM – Open P-TECH**

   o   Free A.I. Education Resources for Students: https://www.ptech.org/open-p-tech/

Chapter 10: Future-Proofing and Unknown Threats

1. **Future of Life Institute**

   o   Addressing Existential Risks from A.I. and Emerging Tech: https://futureoflife.org/

2. **Centre for the Study of Existential Risk (CSER)**

   o   Research on A.I. Safety and Long-Term Threats: https://www.cser.ac.uk/

3. **U.K. Government Office for Science – A.I. and Future Threats**

   o   Reports on Emerging Technology Risks: https://www.gov.uk/government/organisations/government-office-for-science

4. **Singularity University**

   o   Research on Exponential Technologies and A.I.: https://su.org/

Chapter 11: Ethical Human Leadership in A.I. Governance

1. **Human Rights Watch (HRW)**

   o   Global advocacy for human rights, including ethical technology usage: https://www.hrw.org/

2. **Amnesty International**

   o   Human rights advocacy with resources on ethics in technology and the impact of A.I. on rights: https://www.amnesty.org/

3. **Partnership on AI**

   o   Organization focused on the responsible and ethical development of A.I. technology: https://www.partnershiponai.org/

4. **IEEE Global Initiative on Ethics of Autonomous and Intelligent Systems**
   - o Resources on ethically aligned design and ethical leadership in technology: https://ethicsinaction.ieee.org/

5. **Center for Humane Technology**
   - o Non-profit advocating for ethical technology that prioritizes humanity over profit: https://www.humanetech.com/

6. **AI for Humanity (France)**
   - o French government initiative promoting ethical A.I. and social responsibility in technology: https://www.aiforhumanity.fr/

7. **Ethical AI Institute**
   - o Dedicated to advancing ethical A.I. practices and supporting humanity-centered technology development: https://ethical.institute/

8. **United Nations – Artificial Intelligence for Good**
   - o UN initiative exploring A.I. solutions for sustainable development and societal well-being: https://aiforgood.itu.int/

9. **OpenAI Ethics and Public Policy**
   - o OpenAI's initiatives around transparency, ethics, and public policy: https://openai.com/research/

10. **The Leadership Conference on Civil and Human Rights**
    - o Organization that addresses ethical issues in technology impacting civil rights: https://civilrights.org/

11. **The Meta-Center**
    - o Host classes which addresse ethical issues in technology, humanitarian issues, technological challenges and various science, technology, engineering and mathematical principles. https://metacenterchicago.com/

12. **Ethical A.I. Governance**
    - o The author of this work on ethical A.I. governance is well known for his work with artificial intelligence, technology, ethics, and humanitarian pursuits. Private and public organizations and government bodies can benefit from reaching out for person-alized guidance and expertise. Services include **organization-**

al assessments to ensure A.I. systems align with ethical standards, **private classes** on responsible A.I. practices tailored to your team, and **speaking engagements** to inspire and educate audiences on the importance of humanity-centered A.I. governance. By connecting with the author, you gain access to insights that promote responsible A.I. use and safeguard societal values. https://ethicalaigov.com

DEFINITIONS

Algorithmic Accountability: The responsibility of organizations to ensure that algorithms operate transparently, fairly, and without causing harm, especially when they impact decisions about people or communities.

Algorithmic Bias: Systematic and unfair discrimination that can occur when algorithms favor certain groups or outcomes over others, often due to flaws in the data used to train the A.I. or implicit biases embedded in the design.

Algorithmic Manipulation: The intentional adjustment of algorithms by social media or digital platforms to promote, suppress, or control the visibility of certain users or content, often without transparent disclosure to affected parties.

Ambiguity in Enforcement: The practice of applying policies or guidelines inconsistently or selectively, often leading to confusion or perceived unfairness among affected individuals or entities.

Ambiguous Community Guidelines: Loosely defined or vague policies used by platforms to regulate user behavior, often allowing for selective enforcement based on subjective interpretations.

A.I. Ethics: The study and application of moral principles guiding the development and deployment of artificial intelligence to ensure it aligns with human values and societal well-being.

Autonomy: In the context of A.I. and governance, the right of individuals to make informed, independent decisions without undue influence from technology or external entities.

Big Data: Extremely large data sets that can be analyzed computationally to reveal patterns, trends, and associations, often used in A.I. training and decision-making processes.

Censorship: The suppression or prohibition of speech, communication, or information deemed objectionable, harmful, or inconvenient by authority figures or platforms, sometimes through shadowbanning or selective enforcement.

Community Standards: Rules or guidelines set by social media platforms and online communities that govern acceptable behavior and content, which can be ambiguous or inconsistently applied.

Data Privacy: The right of individuals to control their personal data, including how it is collected, used, shared, and stored, particularly in A.I. applications.

Data Transparency: The practice of making data collection, usage, and processing methods clear and accessible to users, ensuring that individuals understand how their data is being utilized.

Disinformation: False or misleading information deliberately spread to deceive people, often amplified through A.I. algorithms or social media platforms.

Exclusionary Practices: Methods or policies that exclude or marginalize certain individuals or groups, often through obstacles like inaccessible customer service, ambiguous guidelines, or biased algorithms.

Exclusionary Terrorism: The use of bureaucratic or systemic methods, such as inaccessible customer service or vague policies, to marginalize or silence individuals or businesses by creating obstacles that deter engagement or resolution of grievances.

Ethical Human Leadership: Leadership that emphasizes responsibility, integrity, transparency, and prioritizing the public good, particularly in contexts where decisions affect large numbers of people.

Ethical Leadership: Leadership that emphasizes responsibility, integrity, transparency, and prioritizing the public good, particularly in contexts where decisions affect large numbers of people.

Financial Manipulation: The act of influencing financial systems, markets, or decisions through data-driven or A.I.-assisted strategies for gain, sometimes in ways that harm others economically.

Financial Terrorism: Actions by corporations or institutions to leverage financial influence or market power in a way that manipulates economic conditions, suppresses competition, or limits economic opportunities for individuals or groups.

Human-Centered Approach: An approach to technology and governance that prioritizes human needs, values, and rights above other considerations, ensuring that A.I. and other innovations benefit society.

Inclusivity: Ensuring that diverse voices, particularly those from marginalized or underrepresented groups, are considered in decision-making processes and that technology is accessible to all.

Influence Algorithms: Algorithms specifically designed to shape or influence user behavior, perceptions, or decisions by promoting certain content or individuals over others.

Manipulative Design: A design approach that subtly or overtly guides users toward specific behaviors, often without their awareness, through mechanisms such as interface design, notifications, or personalized content recommendations.

Misrepresentation: The use of A.I. or data-driven tools to present information or individuals in a misleading way, such as artificially inflating popularity or influence.

Misinformation: False or inaccurate information spread without malicious intent, which can still have harmful consequences, especially when amplified by A.I. algorithms.

Monitoring and Compliance: The process of overseeing adherence to policies, regulations, or guidelines, particularly in A.I. and data-driven applications, to ensure ethical and legal standards are met.

Public Discourse: The open exchange of ideas and information in a public setting, which can be shaped by social media platforms, algorithms, and content moderation practices.

Self-Regulation: The practice of organizations or industries establishing and adhering to their own ethical guidelines or standards without external enforcement, which can be applied to corporations in A.I. development.

Selective Enforcement: The inconsistent or biased application of rules or guidelines, often resulting in certain individuals or groups being unfairly targeted or exempted.

Shadowbanning: A practice where a user's content is intentionally limited in visibility or reach without notifying the user, often used by social media platforms to reduce the influence of certain voices without overt censorship.

Social Terrorism: The use of social platforms, A.I., or big data to manipulate public opinion, control narratives, or restrict the visibility of certain individuals or viewpoints, often without transparency or accountability.

Surveillance Capitalism: A term used to describe a business model where companies collect, analyze, and monetize vast amounts of data from users, often without their explicit consent, for profit.

Synthetic Popularity: The creation of an artificial perception of popularity or influence through algorithms, bots, or other digital tools, making individuals or ideas appear more widely accepted than they are.

Transparency: The quality of being open and honest about policies, practices, and decision-making processes, especially in A.I. governance, so that users understand how their data and experiences are shaped.

Trust and Accountability: The concepts of ensuring that A.I. systems are trustworthy and that their creators are responsible for their impacts, fostering a relationship of confidence between users and technology providers.

User Autonomy: The degree to which users can control their interactions with technology, make informed choices, and avoid manipulation or influence from algorithms.